Understanding Spatio-temporal Variability of Water Resources and the Implications for IWRM in Semi-arid Eastern and Southern Africa

IAHS Scientific Assembly 2017

Proceedings of the International Association of Hydrological Sciences (PIAHS) Volume 378, 2018

Port Elizabeth, South Africa
10 – 14 July 2017

Editors:

H. Makurira
D. Mazvimavi
J.-M. Kileshye-Onema
E. Kapangaziwiri
W. Gumindoga

ISBN: 978-1-5108-6371-2

Printed from e-media with permission by:

Curran Associates, Inc.
57 Morehouse Lane
Red Hook, NY 12571

Some format issues inherent in the e-media version may also appear in this print version.

Printed by Curran Associates, Inc. (2018)

International Association of Hydrological Sciences
Centre for Ecology and Hydrology
Wallingford, Oxfordshire OX10 8BB
United Kingdom

Phone: 441 491 692 442
Fax: 441 491 692 448

http://iahs.info/

Additional copies of this publication are available from:

Curran Associates, Inc.
57 Morehouse Lane
Red Hook, NY 12571 USA
Phone: 845-758-0400
Fax: 845-758-2633
Email: curran@proceedings.com
Web: www.proceedings.com

TABLE OF CONTENTS

https://doi.org/10.5194/piahs-378-1-2018

Open Access
PIAHS

Preface: Understanding spatio-temporal variability of water resources and the implications for IWRM in semi-arid eastern and southern Africa

Hodson Makurira[1], Dominic Mazvimavi[2], Evison Kapangaziwiri[3], Jean-Marie Kileshye Onema[4], and Webster Gumindoga[1]

[1]University of Zimbabwe, P.O. Box MP 167, Mt Pleasant, Harare, Zimbabwe
[2]Department of Earth Sciences, University of the Western Cape, Private Bag X17, Bellville, 7535 Cape Town, South Africa
[3]CSIR NRE, Building 33, Room S222, 627 Meiring Naude Road, Brummeria, Pretoria 0184, South Africa
[4]WaterNet, P.O. Box MP600, Mount Pleasant, Harare, Zimbabwe

Correspondence: Hodson Makurira (makurira@eng.uz.ac.zw)

Published: 29 May 2018

This Special Issue of the Proceedings of the International Hydrological Sciences (IAHS) is a compilation of selected papers that were presented at the IAHS Scientific Assembly held in Port Elizabeth, South Africa, from the 10 to 14 July 2017.

This volume contains 14 peer reviewed papers that focus on integrated water resources management in the southern Africa region and are a culmination of extensive research in five broad themes, namely, (1) hydrological modelling (2) groundwater hydrology (3) application of earth observations in water resources assessments (4) water allocation and use and (5) water quality management. Southern Africa offers wide research opportunities to solve challenges related to the assessment of surface and groundwater resources, water supply and sanitation, water pollution and allocation. However, data scarcity and difficulty in accessing research sites discourage many research efforts. The papers in this Volume are research outputs from mainly postgraduate studies conducted at universities in southern Africa with the sole aim of contributing to knowledge in water sciences that will facilitate informed decision-making and policy formulation by water managers and practitioners in the region.

The editors wish to thank all the contributing authors for their efforts, Waternet which facilitated and funded the bulk of the research that is presented in this volume and the numerous reviewers who contributed to the production of quality papers. Special acknowledgement is also extended to IAHS for facilitating the participation of young scientists at the Scientific Assembly in Port Elizabeth.

Published by Copernicus Publications on behalf of the International Association of Hydrological Sciences.

https://doi.org/10.5194/piahs-378-3-2018

Fuzzy and process modelling of contour ridge water dynamics

Alexander Mhizha[1,2] **and John Ndiritu**[2]

[1]Department of Civil Engineering, University of Zimbabwe, Harare, Zimbabwe
[2]School of Civil and Environmental Engineering, University of Witwatersrand, Johannesburg, South Africa

Correspondence: Alexander Mhizha (amhizha@yahoo.com)

Received: 10 November 2017 – Revised: 29 January 2018 – Accepted: 6 February 2018 – Published: 29 May 2018

Abstract. Contour ridges are an in-situ rainwater harvesting technology developed initially for soil erosion control but are currently also widely promoted for rainwater harvesting. The effectiveness of contour ridges depends on geophysical, hydro-climatic and socio economic factors that are highly varied in time and space. Furthermore, field-scale data on these factors are often unavailable. This together with the complexity of hydrological processes at field scale limits the application of classical distributed process modelling to highly-instrumented experimental fields. This paper presents a framework that combines fuzzy logic and process-based approach for modelling contour ridges for rainwater harvesting where detailed field data are not available. Water balance for a representative contour-ridged field incorporating the water flow processes across the boundaries is integrated with fuzzy logic to incorporate the uncertainties in estimating runoff. The model is tested using data collected during the 2009/2010 and 2010/2011 rainfall seasons from two contour-ridged fields in Zhulube located in the semi-arid parts of Zimbabwe. The model is found to replicate soil moisture in the root zone reasonably well ($NSE = 0.55$ to 0.66 and $PBIAS = -1.3$ to 6.1 %). The results show that combining fuzzy logic and process based approaches can adequately model soil moisture in a contour ridged-field and could help to assess the water dynamics in contour ridged fields.

1 Introduction

In-situ rainwater harvesting techniques are widely promoted in arid and semi-arid regions of the world to mitigate against the effects of dry spells that are frequently experienced in these areas. Some of the in-situ rainwater harvesting techniques that are implemented include planting basins which are found in many parts of sub Saharan Africa (Mupangwa et al., 2012), fanya juus in East African countries mainly in Tanzania and Kenya (Makurira et al., 2010) and contour ridges promoted in Tunisia, Ethiopia (Hengsdijk et al., 2005) and Zimbabwe (Mhizha and Ndiritu, 2013). A contour ridge is an excavated trench with the excavated soil heaped on the down slope side unlike in a fanya juu where the soil is heaped up slope (Fig. 1). The trench provides temporal storage for storm water drained from the field upslope while the ridge prevents the water from flowing downslope once the trench is full.

In Zimbabwe contour ridges were initially adopted and legally enforced for soil erosion control with designs adopted from the United States of America (Elwel, 1981). The initial design comprised of a graded contour ridge constructed along a slope of approximately 5 % to safely drain the water away from a cultivated field. The forced implementation and draining away of water in the dry semi-arid regions was not appreciated by farmers (Hargman, 1996) who later changed the design to dead level contours which are constructed at zero gradient to prevent the storm water from draining away and retain it in the field (Mupangwa et al., 2012). Studies by Mupangwa et al. (2012) showed that soil moisture improvements downslope of a dead level contour ridge were limited to a distance of about 4 m. Mhizha and Ndiritu (2013) further established that soil moisture improvement due to dead level contour ridges was significant in loam soil and not in sandy soil. These observations raise the need for modelling con-

Published by Copernicus Publications on behalf of the International Association of Hydrological Sciences.

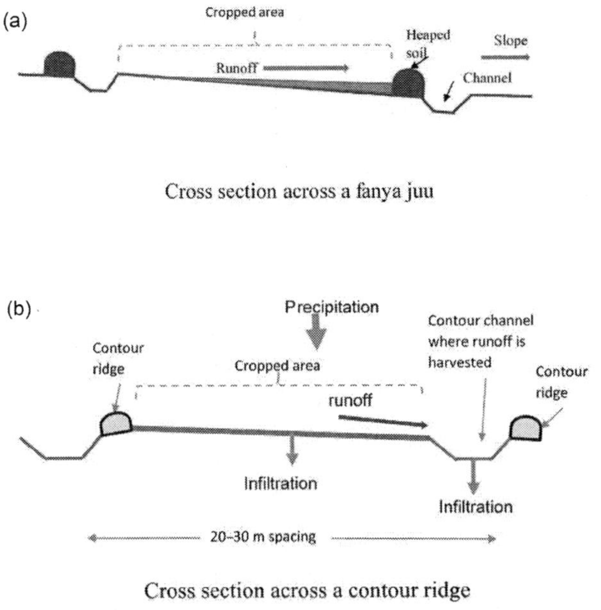

Figure 1. An illustration of the difference between a fanya juu **(a)** and contour ridge **(b)**.

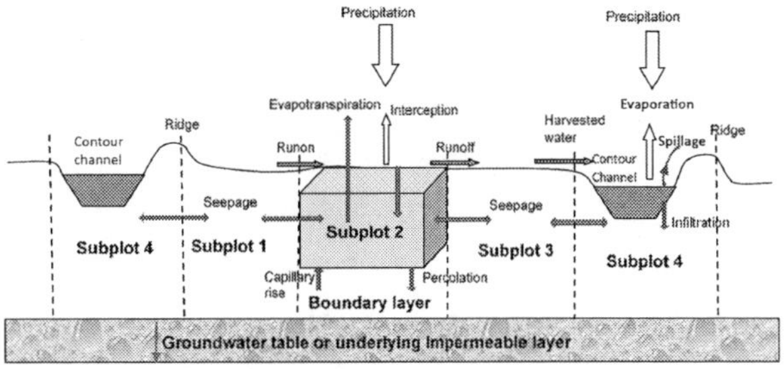

Figure 2. The hydrological system of a contour ridged field.

tour ridged fields to establish water dynamics due to contour ridges.

Modelling of contour ridges, fanya juus and other rainwater harvesting technologies has previously been done using existing models developed for crop water management or for groundwater movement. Hengsdijk et al. (2005) modelled the effect of stone contour ridges (bunds) on crop yield in Ethiopia using the WOFOST (World Food Studies) which is a dynamic crop growth simulation model. The model assumes that runoff reductions due to the stone bunds result in increased infiltration and that this water is stored in the root zone and becomes available to crops. The model does not provide for runoff estimation but rather runoff is obtained from site observations. This implies that uncertain estimates of runoff need to be used on sites where there is no runoff observation. Mwenge Kahinda et al. (2007) used the APSIM (Agricultural Production Systems SIMulator) a crop yield estimation model to model the effect of supplemental irrigation using water from rainwater harvesting systems of rock outcrops and contour ridges in semi-arid areas of Zimbabwe. The HYDRUS model which was developed for modelling groundwater movement was used by Ruidisch et al. (2013) to model water flow in a plastic mulched ridge cultivation system in South Korea. The HYDRUS model does not have provision for estimating runoff and is data intensive limiting its application to experimental sites and not to sites where data can be readily collected by farmers.

Rainwater harvesting modelling requires that results (or models developed with data) from experimental sites are transferred to non-experimental fields for practical application. Parched-Thirst (Young et al., 2002), a process based model was developed to solve such problems of transferability. Such models, whose availability is limited, have a structure that captures the main processes driving rainwater harvesting in order to remain meaningful. Parched-Thirst estimates runoff as infiltration excess using the Green Ampt equation and contour ridges are incorporated as depression storages. Parched-Thirst was successfully used by Magombeyi et al. (2012) to model crop yield from rainwater harvesting by planting basins known locally as Chololo pits. It is suited for this type of rainwater harvesting technology because the spacing between the ridges is small which makes the concept of modifying runoff generation by depression storages justifiable. The spacing of contour ridges that is around 20 m is large enough to warrant that runoff generated in the cropped area be estimated separately. Parched-Thirst does not explicitly estimate the runoff that flows into contour ridges. Soil parameters for the estimation of infiltration are determined using pedo-transfer functions where site data does not exist. This limits its application in analysing contour ridge water dynamics.

This paper presents a model that applies fuzzy logic to estimate runoff in the cropped area and process based approaches for water partitioning in different locations of a contour ridged field. The paper briefly describes the hydrological conceptual framework of a contour ridged field which informs the representative sub regions of a contour ridged field where water partitioning takes place. These sub regions were identified according to hydrological processes that are dominant in each sub region similar to the way sub regions of a representative elementary watershed are identified (Reggiani et al., 1998). The main processes that are dominant in rainwater harvesting by contour ridges are rainfall, interception, runoff, infiltration and unsaturated soil moisture flow.

2 Materials and methods

2.1 Contour ridge design and data collection

The contour ridge layout used in the model was that of the experimental fields set up in Zhulube (Mhizha and Ndiritu, 2013) which were spaced at 20 m and constructed at zero

Table 1. Normalised cluster centre values after data clustering.

(m)	$(c_{0,m})$	(T)	(P_i)	(θ)	(α_N)
1	0.390	0.072	1.000	0.969	1.000
2	0.171	0.242	0.531	0.328	0.666
3	0.412	0.771	0.263	0.734	1.000
4	0.616	0.367	0.735	0.906	1.000
5	0.020	0.572	0.422	0.094	0.010

Key: m is cluster number; $c_{0,m}$ is runoff coefficient for cluster m; T is rainfall duration; P_i is rainfall intensity; θ is soil moisture and α_N is soil type.

gradient to prevent water from leaving the field. Data was obtained from a 3-year long field study of these experimental sites (Mhizha, 2017) and from a previous study in Masvingo (Mugabe et al., 2007). Both sites are located in in the semi-arid areas of Zimbabwe where average rainfall is 540 mm a^{-1}. Data collected comprised of rainfall at daily time step, soil moisture at weekly time step and soil hydraulic parameters. The rainfall data was disaggregated into rainfall intensity at hourly time step using the method developed by Knoesen and Smithers (2009) which is a modification of the rainfall disaggregation model proposed by Boughton (2000). The soil hydraulic parameters collected and used in the modelling are field capacity, the wilting point and infiltration capacity of the soil.

2.2 Modelling water partitioning in a contour ridged field

A contour ridged field comprising of the cropped area, two contour ridges and the root zone (Fig. 2) was considered to be a hydrological system where water partitioning takes place in different components of the system. Water partitioning in these three components of the system were considered for modelling of contour ridge water dynamics. The first water partitioning occurs when rainfall in the cropped area is turned into infiltration, interception and runoff. The runoff generated in the cropped area accumulates in the contour ridge channel where the second partitioning takes place. The accumulated runoff is turned into evaporation, infiltration and if the channel becomes full discharge from the system. The infiltrated water in both the cropped area and the channel add to the soil moisture in the root zone. Within the root zone the third water partitioning occurs. The soil moisture is turned into evapotranspiration (soil evaporation and transpiration), deep percolation and lateral flow to adjacent areas.

To model the water partitioning processes an approach similar to the representative elementary watershed concept (Regianni et al., 1998) was adopted. Each zone of the contour ridge hydrological system where water partitioning occurs was considered a hydrological subzone for which water balance analysis could be done. Water mass flows across the boundaries of each subzone were represented by the main hydrological processes taking place in the water mass balance equations of that subzone.

The model operates on a daily time step which is the time step that farmers can obtain rainfall and evaporation data for their areas. For each time step the model start by executing the rainfall partitioning in the cropped area. Once the model completes water partitioning in the cropped area water partitioning in the contour ridge channel begins. The last water partitioning in the model is carried out in the root zone.

2.2.1 Water balance for rainfall partitioning

The inputs in the water balance equation for rainfall partitioning are rainfall, interception and runoff while infiltration is taken as the residual of the water balance equation. At a time step of one day the change in surface water storage for a cropped area is assumed to be zero thus leaving infiltration as the only unknown if interception and runoff can be independently estimated. Interception reduces the amount of rainfall that reaches the soil and become available to other hydrological processes of runoff and infiltration. The model estimates interception using a threshold-based approach and includes all evaporative demands that feed moisture back into the atmosphere (de Groen and Savenije, 2006). In this approach, when rainfall stops, interception storage is reduced by evaporation until there is no more storage or until the evaporative demand for the day is achieved (Makurira et al., 2010). Runoff is estimated using a fuzzy logic based method. This method uses a set of linear regression models where rainfall intensity, rainfall duration, soil moisture (for the previous day) and soil type are input data and runoff is the output. This method is described in more detail in Sect. 2.2. The cropped area is divided into subplots to allow accumulation of runoff downslope of a contour ridge. Thus effective precipitation at each subplot takes into account runoff from upslope subplots. This ensures that downslope subplots generate more runoff than upslope subplots. Finally the subplot upslope of the contour ridge channel discharges water into the channel thus making it input into the runoff partitioning subzone. Equation (1) describes the water balance for rainfall partitioning.

$$\frac{\Delta \varsigma}{\Delta t} = e^{\mathrm{rz}} + e^{\mathrm{atm}} + e^{\mathrm{ada}} + e^{\mathrm{crc}} \tag{1}$$

Where: $\Delta \varsigma / \Delta t$ is change in surface storage; e^{atm} is exchange with atmosphere (precipitation and interception or evaporation); e^{ada} is net exchange of runoff with adjacent subplot surface area; e^{crc} is exchange with the contour ridge channel (runoff); and e^{rz} is exchange with the root zone (infiltration into the subplot).

2.2.2 Water balance for runoff partitioning

The inputs for runoff partitioning are rainfall as data for the site, runoff from the rainfall partitioning and evaporation. Infiltration in the contour ridge channel is determined from the

Table 2. Model consequent coefficients from shuffled evolution calibration.

Sub model Number	Constant ($c_{0,m}$)	rain duration ($c_{1,m}$)	rain intensity ($c_{2,m}$)	soil moisture ($c_{3,m}$)	soil type ($c_{4,m}$)
1	0.042	0.107	0.163	0.129	0.089
2	−0.244	0.266	0.280	0.238	0.187
3	−0.021	0.181	0.181	0.136	0.144
4	0.027	0.270	0.059	0.208	0.271
5	−0.008	0.020	0.020	0.021	0.304

Table 3. Comparison of the model sensitivity to selected parameters.

Model Parameter	minimum CN	maximum CN	deltaCN
kpan	−100.7	−1.6	99.1
soilevafactor	−4.5	−1	3.5
reduction scale	−64.1	37	101.1
pfraction	−4.6	−0.8	3.8
fcapacity	−50	5.5	55.5
wiltpoint	−0.3	−0.1	0.2
infilcapacity	0	0	0

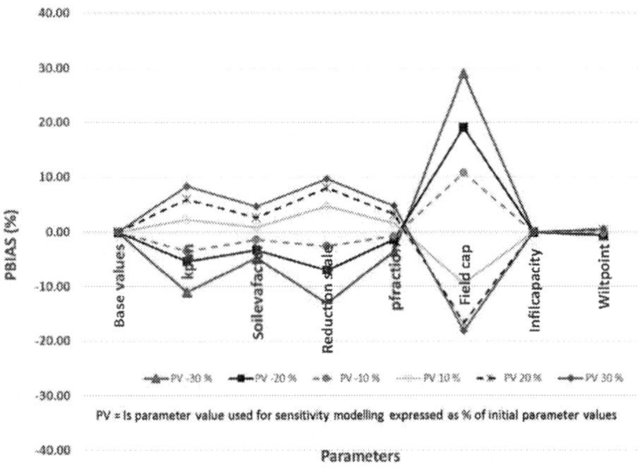

Figure 3. Results of model sensitivity analysis.

infiltration capacity of the soil and water availability in the contour ridge channel. The water balance allows for spillage if the net increase in volume leads to a value exceeding the storage capacity of the contour ridge channel. Equation (2) describes the runoff partitioning water balance.

$$\frac{\Delta \check{C}}{\Delta t} = e^{\mathrm{crz}} + e^{\mathrm{atm}} + e^{\mathrm{ca}} + e^{\mathrm{ad}} \qquad (2)$$

Where: $\Delta \check{C}/\Delta t$ is change in contour ridge channel storage; e^{crz} is exchange with channel root zone (infiltration); e^{atm} is exchange with atmosphere (precipitation and evaporation); e^{ca} is exchange with cropped area (run-on into contour ridge channel); e^{ad} is exchange with adjacent area (spillage).

2.2.3 Water balance for soil moisture partitioning

Like in the rainfall partitioning the contour ridged field is divided into subzones for soil moisture partitioning and mass water balance is carried out separately in each subzone. Infiltration estimated from the rainfall partitioning and the runoff partitioning enters the root zone as input for the soil moisture partitioning zone.

Evapotranspiration includes both evaporation from soil and transpiration by the crops and is estimated by considering evaporation and transpiration processes separately in terms of the FAO-56 method as applied by Makurira et al. (2010) and Allen (2000). The evaporation component is estimated using equation 3 and includes a reduction scale parameter which needs to be calibrated (Makurira et al., 2010).

$$E_{\mathrm{s}} = \max\left(1 - I_{\mathrm{LA}}, 0\right) \max\left(k_{\mathrm{s}} k_{\mathrm{p}} E_{o} - E_{\mathrm{I}}, 0\right) f_{\mathrm{ms}} \qquad (3)$$

Where: E_{s} is soil evaporation (mm day^{-1}); E_o is pan evaporation for the day, I_{LA} is the leaf area index; E_{I} is evaporation from interception which has an effect of reducing evaporation demand in the atmosphere; k_{s} is soil evaporation factor (equivalent to crop factor in cropped areas); k_{p} is the pan evaporation factor and f_{ms} is moisture stress reduction factor given by Eq. (4).

$$f_{\mathrm{ms}} = \min(\exp((\upsilon_t - \upsilon_{\mathrm{sat}})/b), 1) \qquad (4)$$

Where: υ_{sat} is maximum soil moisture (which is soil moisture at saturation) in the root zone (mm) and b is a reduction scale (mm).

The vertical and lateral soil moisture fluxes are estimated using Eq. (5). The hydraulic conductivity and diffusivity (K_{s} and F_z in Eq. 5) vary hysterically with moisture. In order to cater for hysteresis in hydraulic conductivity Mahrt and Pan (1984) proposed a thin layer soil hydrology model in which the hydraulic conductivity of a layer of soil when drying would be based on the soil moisture of the lower layer and when wetting the hydraulic conductivity would be based on the soil moisture of the upper layer. This approach was adopted here.

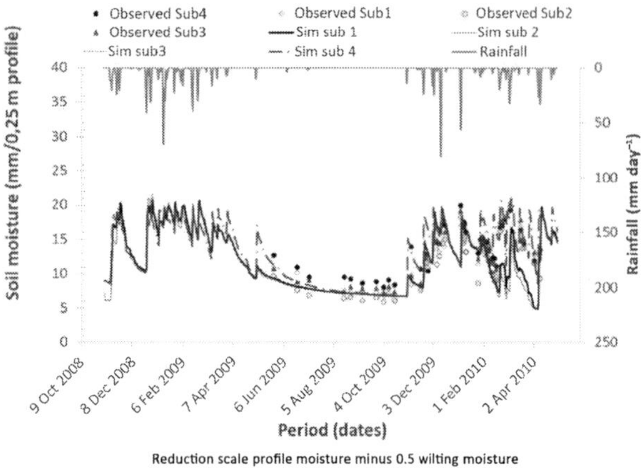

Figure 4. Model performance with reduction scale based on prevailing soil moisture.

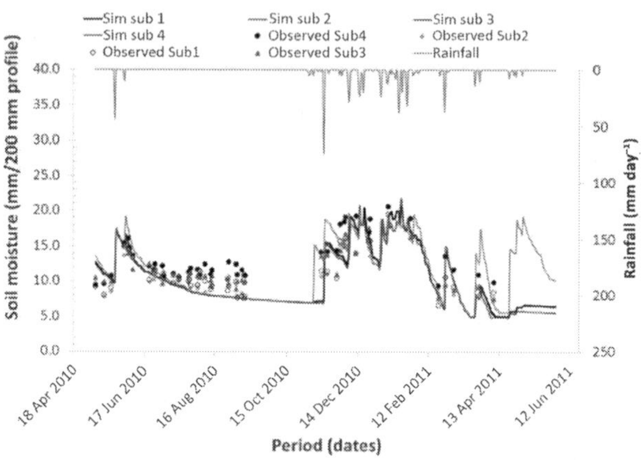

Figure 5. Model verification farm A.

$$\Phi_{m,t} = K_s F_z \frac{\partial \theta}{\partial x} \quad (5)$$

Where: $\Phi_{m,t}$ is soil moisture flux (flow) into or out of subplot m at time t; $\partial\theta/\partial x$ is the soil moisture gradient given by Eq. (6) and K_s F_z is hydraulic conductivity and diffusivity of the soil.

$$\frac{\partial \theta}{\partial x} = \frac{\theta_{m,t} - \theta_{m+1,t-1}}{\Delta x} \quad (6)$$

Where: $\theta_{m,t-1}$ and $\theta_{m+1,t-1}$ are the soil moisture in subplot m and $m+1$ at end of time t and $t-1$, respectively and Δx is distance between the centre of subplots m_{d} and $m_{\mathrm{d}+1}$.

The water transmission and storage processes were modelled and incorporated in the subsurface water balance. Mhizha (2017) provides more details of how this was done. Equation (7) describes the soil moisture zone water balance.

$$\frac{\Delta \theta}{\Delta t} = e^{\mathrm{ca}} + e^{\mathrm{atm}} + e^{\mathrm{as}} + e^{\mathrm{gw}} + e^{\mathrm{mp}} \quad (7)$$

Where: $\Delta\theta/\Delta t$ is change in soil moisture during the time step; $\theta_{m,t}$ is soil moisture in subzone m at time t (current time); $\theta_{m,t-}$is soil moisture in subzone m, at time $t-1$; e^{ca} is exchange with cropped area surface (infiltration); e^{atm} is exchange with the atmosphere (evapotranspiration); e^{as} is exchange with adjacent subzone (seepage); e^{gw} is exchange with groundwater system (percolation); e^{mp} is exchange with macropore spaces within the same subplot.

2.3 Fuzzy rainfall runoff modelling

This method makes use of several linear regression models for estimating runoff amounts as a function of rainfall intensity, rainfall duration, soil moisture and soil type. Application of linear regression models for estimating runoff at field scale is common for experimental sites (Li et al., 2004; Walker and Tsubo, 2003). To try and ensure a balanced influence of the independent variables on modelled runoff, all the data were normalized for the fuzzy modelling. Fuzzy modelling typically follows the IF-(antecedent part)–THEN-(consequence part) approach which is controlled by an inference system developed through the Mamdani or the Tagagi-Sugeno system (Katambara and Ndiritu, 2009). By applying the Tagagi-Sugeno fuzzy model development approach 5 multiple linear regression models of the form given in Eq. (8) were established. The linear modelling approach in Eq. (8) is in line with the regression models normally used to model runoff at field scale (Li et al., 2004; Walker and Tsubo, 2003). Each of these models have their focal point whose positions are given in Table 1. Fuzzification of data points was achieved through establishing the degree to which each data set belonged to any of the sub models. This together with the computation of the overall runoff coefficient from the partial submodel contributions (defuzzification) were done following the method described in Katambara and Ndiritu (2009) for a fuzzy inference system developed to model stream flow.

$$q_{j,m} = c_{0,m} + c_{1,m} P_{ij} + c_{2,m} T_j + c_{3,m}\theta_j + c_{4,m}\alpha_{\mathrm{N}j} \quad (8)$$

Where: $q_{j,m}$ is normalised runoff coefficient (mm) contributed by partial model m during rainfall event j; $P_{i,j}$ is normalised rainfall intensity during rainfall event j (mm); T_j is normalised rainfall duration for rainfall event j (hours); θ_j is normalised root zone soil moisture during event j (mm); $\alpha_{\mathrm{N}j}$ is normalised soil parameter defining soil hydraulic conductivity for event j and $c_{0,m}$, $c_{1,m}$, $c_{2,m}$, $c_{3,m}$ and $c_{4,m}$ are coefficients to input variables for cluster m.

2.4 Sensitivity analysis, model calibration and performance assessment

A sensitivity analysis was carried out to establish data and model parameters that affect model outputs. This involved

Table 4. Model performance during calibration and verification stages.

MPC	MAC	MVA	MVB
NSE	0.657	0.522	0.622
PBIAS	−1.28	6.1	−5.3
R^2	0.586	0.627	0.615

Key: MPC is model performance criteria; MAC is model performance after calibration for farm A; MV model verification (A-farm A and B-farm B). NSE is Nash-Sutclife efficiency; PBIAS is Percent Bias and R^2 is Coefficient of determination.

modifying selected input parameters while holding all other parameters constant (Hoyos and Cavalcante, 2015; Kumar et al., 2014). The sensitivity of model output to the different parameters was compared using the condition number which expresses the rate of change of the dependent variable with respect to the rate of change in the independent variables (Hoyos and Cavalcante, 2015). The parameters that were assessed for sensitivity are shown in Table 3. The values were increased or reduced by 0 to 30 % of base value.

Model calibration for the fuzzy component applied the SCE-UA method (Duan et al., 1992) with the objective of minimizing the root mean square error between the simulated and observed runoff. The calibrated coefficients of each sub model are given in Table 2. The overall model calibration was based on the sensitivity analysis results after which the model was calibrated by adjusting the value of the sensitive parameter.

Performance of the overall model was assessed using the Nash-Sutcliffe efficiency (NSE) (Eq. 9), the Percent bias (PBIAS) determined by Eq. (10) and coefficient of determination (R^2) (Eq. 11) as defined by Moriasi et al. (2007).

$$\mathrm{NSE} = 1 - \left[\frac{\sum_{i=1}^{n} \left(Y_{i,\mathrm{obs}} - Y_{i,\mathrm{sim}}\right)^2}{\sum_{i=1}^{n} \left(Y_{i,\mathrm{obs}} - \overline{Y_{\mathrm{obs}}}\right)^2} \right] \tag{9}$$

$$\mathrm{PBIAS} = \left[\frac{\sum_{i=1}^{n} \left(Y_{i,\mathrm{obs}} - Y_{i,\mathrm{sim}}\right) \times (100)}{\sum_{i=1}^{n} Y_{i,\mathrm{obs}}} \right] \tag{10}$$

$$R^2 = \left[\frac{\sum_{i=1}^{n} \left(Y_{i,\mathrm{obs}} - \overline{Y_{\mathrm{obs}}}\right)\left(Y_{i,\mathrm{sim}} - \overline{Y_{\mathrm{sim}}}\right)}{\sqrt{\sum_{i=1}^{n} \left(Y_{i,\mathrm{obs}} - \overline{Y_{\mathrm{obs}}}\right)^2} \sqrt{\sum_{i=1}^{n} \left(Y_{i,\mathrm{sim}} - \overline{Y_{\mathrm{sim}}}\right)^2}} \right]^2 \tag{11}$$

Where: $Y_{i,\mathrm{obs}}$ and $Y_{i,\mathrm{sim}}$ are the observed and simulated values for data point i and $\overline{Y_{\mathrm{obs}}}$ and $\overline{Y_{\mathrm{sim}}}$ are observed and simulated mean values for all the data points.

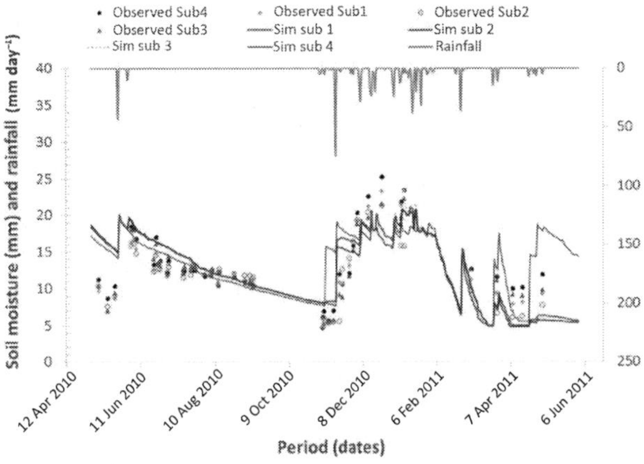

Figure 6. Model verification farm B.

The data that was used for verification was for a rainfall season different from that which was used during model calibration. Data from field A for the season 2009/2010 were used for model calibration while that from field A and field B for the season 2010/2011 were used to show how the model would perform in an independent field not used to calibrate the model.

3 Results and discussion

Figure 5 shows the PBIAS for eight parameters that were used in the sensitivity analysis. The results showed that the model is sensitive to the value of the field capacity, the reduction scale, pan evaporation data and soil evaporation. This was further confirmed by the condition number (CN) as shown in Table 3. The condition number was calculated by considering the change in simulated soil moisture when a parameter value is changed by the indicated percentage compared to the simulated value when all parameters maintain their base values as the dependent variable and the percentage change in parameter value as the independent variable (Hoyos and Cavalcante, 2015).

Except the reduction scale the other three (field capacity, pan evaporation and soil evaporation) values can be improved through more accurate field observations. Therefore the reduction scale was established to be the only parameter that required calibration. The reduction scale was also found to be the next most sensitive parameter after field capacity of the soil. Trial modelling runs also revealed that reduction scale was dependent on the prevailing soil moisture (Eq. 4) although Makurira et al. (2010) had previously used a constant reduction scale (Fig. 4). The value of the reduction scale after calibration was found to be equal to the prevailing soil moisture minus the soil moisture at wilting point. This shows that the reduction scale is not constant but a variable parameter that depends on soil moisture. This conforms to the units of the reduction scale as given in Eq. (4).

The uncertainty in the modelled soil moisture resulted largely from the uncertainty in field data (rainfall, evaporation and soil parameters) and the model structure. The structure of the model is such that runoff modelled by the fuzzy component of the model becomes input data to the process based component of the model. Zere et al. (2005) established a performance level of $R^2 = 0.49$ for a maize plot and $R^2 = 0.59$ for a bare plot using the PutRun model. In addition to the uncertainty arising from the modelled runoff soil moisture simulated in one time step is used as input data when modelling the next time step. These two factors affect the overall performance of the model.

Results of the verification of model performance are shown in Figs. 5 and 6. The model performance (Table 4) with NSE value of 0.52 to 0.66 and PBias value of -1.3 % to 6.1 for soil moisture compare favourably with the results obtained by Ruidisch et al. (2013) who modelled ridge cultivation in South Korea using HYDRUS 2-D model on a similar soil (loam) and obtained a NSE value of 0.48 and PBIAS of 12 %. Again the R^2 value of 0.59 to 0.63 for the modelled soil moisture indicates fairly good model performance considering that uncertainty in estimating runoff affects overall model output.

The strength of this modelling approach lies in the fact that it can be applied to a field where data on runoff does not exist. The runoff of such a field with limited data is estimated by taking into account the prevailing conditions on the field and relates them to the focal points of the fuzzy sub models developed from fields that had data. This is similar to the approach that is taken when modelling ungauged sites where model parameters from gauged basins are extrapolated to ungauged sites (Kapangaziwiri et al., 2009). This means that there is potential to apply this approach in data scarce situations.

4 Conclusions

A realistic modelling of contour ridged field water dynamics was achieved using fuzzy logic rainfall-runoff modelling and process based approaches of quantifying soil moisture fluxes. The model was tested using field data collected from a site in rural Zimbabwe. The model performed reasonably well in calibration with a Nash-Sutcliffe Efficiency of 0.657, overall percent bias of -1.28 % and a Coefficient of determination of 0.586. In verification the model performed equally well with Nash-Sutcliffe Efficiency of 0.522, overall percent bias of 6.1 % and a Coefficient of determination of 0.627 on the loam soil and Nash-Sutcliffe Efficiency of 0.622, overall percent bias of -5.3 % and a Coefficient of determination of 0.615 on the sandy soil. The applicability of this modelling to contour ridge design for rainwater harvesting needs to be assessed further. The modelling approach applied here could also be applicable to hydrological modelling in data-scarce situations.

Data availability. The model code can be obtained from the University of Witwatersrand Library or from the authors. The underlying research data set is available in the Supplement.

The Supplement related to this article is available online at https://doi.org/10.5194/piahs-378-3-2018-supplement.

Competing interests. The authors declare that they have no conflict of interest.

Special issue statement. This article is part of the special issue "Understanding spatio-temporal variability of water resources and the implications for IWRM in semi-arid eastern and southern Africa". It is a result of the IAHS Scientific Assembly 2017, Port Elizabeth, South Africa, 10–14 July 2017.

Acknowledgements. The authors acknowledge and thank the following organizations and persons for their assistance in this study: The Challenge Programme for Water and Food and The University of Witwatersrand for providing financial support for this study which was part of the PhD study for Alexander Mhizha. Francis Themba Mugabe for providing data for the sites in Masvingo Province, Zimbabwe. Lewis Ndlovu and Sithabile Ndlovu for assistance in field data collection in Zhulube Catchment, Matebeleland South Province, Zimbabwe.

Edited by: Hodson Makurira
Reviewed by: Raphael Tshimanga and one anonymous referee

References

Allen, R. G.: Using the FAO-56 dual crop coefficient method over an irrigated region as part of an evapotranspiration intercomparison study, J. Hydrol., 229, 27–41, 2000.

Boughton, W.: A model for disaggregating daily to hourly rainfalls for design flood estimation, Report 00/15, Cooperative Research Centre for Catchment Hydrology, Monash University, Clayton, Victoria, Australia, 2000.

de Groen M. M. and Savenije H. H. G.: A monthly interception equation based on the statistical characteristics of daily rainfall, Water Resour. Res., 42, W2417 to W2427, https://doi.org/10.1029/2006WR005013, 2006.

Duan, Q. Y., Sorooshian, S., and Gupta, V.: Effective and efficient global optimization for conceptual rainfall-runoff models, Water Resour. Res., 28, 1015–1031, 1992.

Elwell, H. A.: Contour Layout Design, Department of Conservation and Extension, Government Printers, Harare, Zimbabwe, 1981.

Hagmann, J.: Mechanical soil conservation with contour ridges: cure for, or cause of, rill erosion?, Land Degrad. Dev., 7, 145–160, 1996.

Hengsdijk, H., Meijerink, G. W., and Mosugu, M. E.: Modelling the effect of three soil and water conservation practices in Tigray, Ethiopia, Agr. Ecosyst. Environ., 105, 29–40, 2005.

Hoyos, E. M. and Cavalcante, A. L. B.: Sensitivity Analysis of One-Dimensional Infiltration Models, EJGA, 20, 4313–4324, 2015.

Kapangaziwiri, E., Hughes, D. A., and Wagener, T.: Towards the development of a consistent uncertainty framework for hydrological predictions in South Africa, New Approaches to Hydrological Prediction in Data-sparse Regions (Proc. of Symposium HS.2 at the Joint IAHS & IAH Convention, Hyderabad, India, September 2009), IAHS-AISH P., 333, 84–93, 2009.

Katambara, Z. and Ndiritu, J.: A fuzzy inference system for modelling streamflow: Case of Letaba River, South Africa, Phys. Chem. Earth, 34, 688–700, 2009.

Knoesen, D. and Smithers, J.: The development and assessment of a daily rainfall disaggregation model for South Africa, Hydrolog. Sci. J., 54, 217–233, doi:10.1623/hysj.54.2.217, 2009.

Kumar, R., Shankar, V., and Jat, M. K.: Sensitivity Analysis of Nonlinear Model Parameters in a Multilayer Root Zone, Journal of Hydrologic Engineering, ASCE, 19, 462–471, 2014.

Li, X.-Y., Xie, Z.-K., and Yan, X.-K.: Runoff characteristics of artificial catchment materials for rainwater harvesting in the semiarid regions of China, Agr. Water Manage., 65, 211–224, 2004.

Magombeyi, M. S., Morardet, S., Taigbenu, A. E., and Cheron, C.: Food insecurity of smallholder farming systems in B72A catchment in the Olifants River Basin, South Africa, Afr. J. Agric. Res., 7, 278–297, 2012.

Mahrt, L. and Pan, H.: A two-layer model of soil hydrology, Bound.-Lay. Meteorol., 29, 1–20, 1984.

Makurira, H., Savenije, H. H. G., and Uhlenbrook, S.: Modelling field scale water partitioning using on-site observations in sub-Saharan rainfed agriculture, Hydrol. Earth Syst. Sci., 14, 627–638, https://doi.org/10.5194/hess-14-627-2010, 2010.

Mhizha, A.: Contour ridge modelling using fuzzy logic and process based approaches for improved rainwater harvesting, unpublished PhD Thesis, University of Witwatersrand, 2017.

Mhizha, A. and Ndiritu, J. G.: Assessing crop yield benefits from in-situ rainwater harvesting through contour ridges in Semi-Arid Zimbabwe, Physics and Chemistry of the Earth, Elsevier, 2013.

Moriasi, D. N., Arnold, J. G., Van Liew, M. W., Bingner, R. L., Harmel, R. D., and Veith, T. L.: Model evaluation guidelines for systematic quantification of accuracy in watershed simulations, Transactions of the ASABE, American Society of Agricultural and Biological Engineers ISSN 0001-2351, 50, 885–900, 2007.

Mugabe, F. T., Hodnett, M. G., and Senzanje, A.: Effect of temporal rainfall distribution and soil type on soil moisture and runoff generation in semi-arid Zimbabwe, Water Policy, 38, 1–16, 2007.

Mupangwa W., Twomlow S. J., and Walker S.: Dead level contours and infiltration pits for risk mitigation in smallholder cropping systems of southern Zimbabwe, Phys. Chem. Earth Pt A/B/C, 47, 166–172, 2012.

Mwenge Kahinda, J., Rockstrom, J., Taigbenu, A. E., and Dimes, J.: Rainwater harvesting to enhance water productivity of rainfed agriculture in the semi-arid Zimbabwe, Phys. Chem. Earth, 32, 1068–1073, 2007.

Reggiani, P., Sivapalan, M., and Hassanizadeh, S. M.: A unifying framework for watershed thermodynamics: balance equations for mass, momentum, energy and entropy, and the second law of thermodynamics, Adv. Water Resour., 22, 367–398, 1998.

Ruidisch, M., Kettering, J., Arnhold, S., and Huwe, B.: Modeling water flow in a plastic mulched ridge cultivation system on hillslopes affected by South Korean summer monsoon, Agr. Water Manage., 116, 204–217, 2013.

Walker, S. and Tsubo, M.: PutuRun: A simulator for rainfall-runoff-yield processes with in-field water harvesting, WRC Report No. K8/486, Water Research Commission Pretoria, 2003.

Young, M. D. B, Gowing, J. W., Wysecure, G. C. L., and Hatibu, N.: Patched-Thirst: Development and validation of a process-based model of rainwater harvesting, Agr. Water Manage., 55, 121–140, 2002.

Zere, T. B., van Huyssteen, C. W., and Hensley, M.: Estimation of runoff at Glen in the Free State Province of South Africa, Water SA, 31, available at: https://www.ajol.info/index.php/wsa/article/view/5116/12728 (last access: 14 March 2018), 2005.

https://doi.org/10.5194/piahs-378-11-2018

Parameter and input data uncertainty estimation for the assessment of water resources in two sub-basins of the Limpopo River Basin

Nadia Oosthuizen[1,2], **Denis A. Hughes**[2], **Evison Kapangaziwiri**[1], **Jean-Marc Mwenge Kahinda**[1], and **Vuyelwa Mvandaba**[1,2]

[1]CSIR, Natural Resources and Environment, P.O. Box 395, Pretoria 0001, South Africa
[2]Institute for Water Research, Rhodes University, P Bag 94, Grahamstown, 6140, South Africa

Correspondence: Nadia Oosthuizen (nadia.oosthuizen.geo@gmail.com)

Received: 10 November 2017 – Revised: 23 February 2018 – Accepted: 5 March 2018 – Published: 29 May 2018

Abstract. The demand for water resources is rapidly growing, placing more strain on access to water and its management. In order to appropriately manage water resources, there is a need to accurately quantify available water resources. Unfortunately, the data required for such assessment are frequently far from sufficient in terms of availability and quality, especially in southern Africa. In this study, the uncertainty related to the estimation of water resources of two sub-basins of the Limpopo River Basin – the Mogalakwena in South Africa and the Shashe shared between Botswana and Zimbabwe – is assessed. Input data (and model parameters) are significant sources of uncertainty that should be quantified. In southern Africa water use data are among the most unreliable sources of model input data because available databases generally consist of only licensed information and actual use is generally unknown. The study assesses how these uncertainties impact the estimation of surface water resources of the sub-basins. Data on farm reservoirs and irrigated areas from various sources were collected and used to run the model. Many farm dams and large irrigation areas are located in the upper parts of the Mogalakwena sub-basin. Results indicate that water use uncertainty is small. Nevertheless, the medium to low flows are clearly impacted. The simulated mean monthly flows at the outlet of the Mogalakwena sub-basin were between 22.62 and 24.68 Mm^3 per month when incorporating only the uncertainty related to the main physical runoff generating parameters. The range of total predictive uncertainty of the model increased to between 22.15 and 24.99 Mm^3 when water use data such as small farm and large reservoirs and irrigation were included. For the Shashe sub-basin incorporating only uncertainty related to the main runoff parameters resulted in mean monthly flows between 11.66 and 14.54 Mm^3. The range of predictive uncertainty changed to between 11.66 and 17.72 Mm^3 after the uncertainty in water use information was added.

1 Introduction

The continued socio-economic development of riparian countries of the Limpopo River leads to pressure on the water resources of the basin. The management of the water resources is, therefore, critical to avoid conflict and ensure equity and accessibility to the resource for both urban and rural populations. There are also various other competing water users such as the environment (environmental water requirements), livestock farming, irrigation, and mining operations. An identification of runoff generating processes, water use and a clear understanding of their linkages within the basin are needed for improved quantification of the water resources of the basin, a requisite for better resource management. There are several challenges within the Limpopo, including shortages of water (e.g. droughts in the southern part of Zimbabwe and western Botswana), flooding (especially in the Mozambique part of the basin) and water quality issues (e.g. in the Oliphant's sub-basin in South Africa). One way of contributing to finding solutions for some of these problems is the understanding of the process linkages for improved resource quantification. However, in the absence of

Published by Copernicus Publications on behalf of the International Association of Hydrological Sciences.

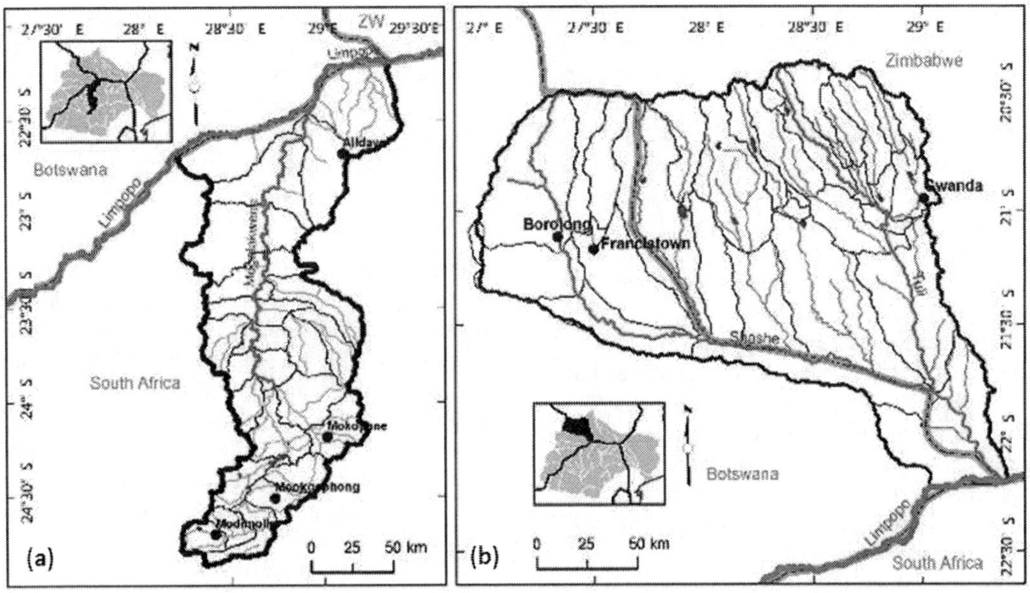

Figure 1. Map of the Mogalakwena **(a)** and Shashe **(b)** sub-basins and their locations in the Limpopo River Basin.

historical observed data (large parts of the basin are virtually ungauged) of the different aspects of the hydrology of the basin (such as rainfall, streamflow, water use, etc.) hydrological models are generally used to generate data and information that inform management and policy decision making.

Ideally, any hydrological model should be based on a sound conceptual understanding of the processes operating in the basin and should be backed by quantitative information that can be used for the parameterization of the model (Hughes et al., 2006, 2010). However, these data are often inadequate in many sub-basins, necessitating the incorporation of the uncertainty related to the estimation process. Given the diversity of the four riparian countries on data collection and management, the quantity and quality of the data in the basin varies and therefore a framework that incorporates estimates of uncertainty should be applied to deal with this challenge. Earlier studies (e.g. Matji and Görgens, 2001) have generally concentrated on the main stem of the basin and this study explains the possible uncertainty related to the modelling process.

2 Study Area

Two physically and socio-economically contrasting sub-basins were chosen for the study (Fig. 1). The Mogalakwena River rises as the Nyl River south of Mokopane flows northwards into the Limpopo (Busari, 2008). The Mogalakwena sub-basin, with a drainage area of 19 400 km^2, is located in South Africa and is very densely populated and industrialized incorporating the towns of Modimolle, Mookgopong, and Mokopane (Fig. 1a). The Shashe sub-basin (Fig. 1b), shared between Botswana and Zimbabwe, rises on the border of the two countries. It flows south, past Francistown (in Botswana) where after the flow changes to a south-easterly direction along the border for about 362 km until it flows into the Limpopo River.

While the sub-basins are near one another, they however have different physical characteristics such as rainfall, evaporation, soils, geology and land cover which, in turn, influence the runoff rates, water use, and abstractions. Also, since the catchments are located in different countries, the data collection methods vary affecting the data quality and availability. In South Africa, hydrological data can be accessed freely from sources such as the Water Resources Studies (1990, 2005 and 2012), national departments such as the Department of Water and Sanitation (DWS) and the Department of Agriculture, Forestry and Fisheries (DAFF), whereas hydrological data for Botswana and Zimbabwe are more difficult to access, the records of poorer quality often with missing values, and large areas are ungauged. The two study areas are therefore chosen to show the impacts of different physical and hydro-climatic data conditions on estimated water resources.

3 Data and Methodology

The Pitman model has been widely used for water resources assessment in most of the southern Africa region since its development in the early 1970s (Wilk and Hughes, 2002). The Spatial and Time Series Information Modelling (SPATSIM, Hughes and Forsyth, 2006) version of the Pitman model (Pitman, 1973) was used to quantify water resources of the selected sub-basins. It is a conceptual type model with parameters that are associated with components that represent the main hydrological processes (and human impacts) that operate at a sub-basin scale (Hughes et al., 2010). The SPATSIM version of the model is designed to handle multiple model runs that are required to consider uncertainty issues arising from the quantity and quality of the input data (Kapangaziwiri et al., 2012). This version also has explicit surface-ground water interaction routines (Hughes, 2004) and a wetland function. Table 1 shows the main parameters of the model (Hughes et al., 2006).

Runoff is mainly generated by two model functions. The first is an asymmetrical triangular distribution of catchment absorption rates defined by parameters ZMIN, ZAVE and ZMAX. The second function determines the drainage rate from the main moisture storage (S, with a capacity of ST, mm). This storage is depleted by evapotranspiration, interflow and groundwater recharge. The maximum interflow (FT, mm month^{-1}) and recharge (GW, mm month^{-1}) rates occur at ST, while two power functions (parameters POW and GPOW) determine these rates at lower values of moisture storage (S, mm). Recharge is routed through a groundwater storage function that accounts for evapotranspiration losses, drainage to other catchments and contributions to base flow.

The model also has functions that simulate the impact of human activities like small farm dams, large reservoirs and irrigated agriculture in managed basins. Detailed descriptions of the model can be found in Hughes et al. (2006) and Kapangaziwiri et al. (2012).

3.1 Data collection

For this study, the impact of uncertainty related to farm dam and irrigation data was evaluated. The data sources that were used include: the 2012 national water resources assessment study of South Africa (called the WR2012) database, data from the South African Department of Agriculture, Forestry and Fisheries (DAFF) and remote sensing products for the Mogalakwena sub-basin. For Shashe, besides remotely-sensed information, the data from the Limpopo River Basin Monograph study (LIMCOM, 2013) were used.

The farm dam data (and all the other data used in this study) were collected at the scale at which water resources decisions are made (i.e. at the so-called quaternary catchment for the Mogalakwena and at the sub-zone scale for Shashe). In this study, both quaternary and sub-zones are referred to just as catchment. For purposes of simplicity and also based on the model used, the individual small farm dams in each catchment were added up to form one dam at the outlet, whose parameters (e.g. full supply capacity and area) were then subsequently specified and the level of uncertainty where necessary. Only the surface areas of the identified farm dams were estimated from the remote sensing and through a manual digitizing process. The dam capacities had to be calculated using a generalised relationship between capacity and area at full supply level provided by LIMCOM (2013). While many approaches to the estimation of dam capacities could be used (e.g. Sayl et al., 2016; Hughes and Mantel, 2010 and Sawunyama et al., 2006) this equation was chosen since it was applied successfully in the whole of the Limpopo basin, and is given as:

$$\text{Surface Area} = 0.4 \cdot (\text{Capacity})^{0.7} \quad (1)$$

The calculated volumes were similar to those contain in existing national databases and were therefore assumed adequate to use. For running the uncertainty in the SPATSIM version of the Pitman model, minimum (Min) and maximum (Max) values are required in the model setup. The percentage differences between the average of minimum and maximum of dam volumes or irrigation areas was thus determined to represent the uncertainty. This is given by: [maximum average − minimum average)/maximum average value] × 100 (Sawunyama et al., 2011).

WR2012 irrigation data for Mogalakwena sub-basin were compared to the data received from the South African Department of Agriculture, Forestry and Fisheries (DAFF) and remotely-sensed data. For the Shashe sub-basin, the LIMCOM (2013) data were also compared to areas digitised in Google Earth.

3.2 Uncertainty analysis

The analysis focused on the impact of uncertainties related to the main physical parameters of the runoff generation process and water use data (i.e. irrigation and farm dams) on the estimation of sub-basin water resources. Parameter uncertainties were considered first, before the incorporation of water use uncertainties.

In this study, the parameter values for each run of the model are independently randomly sampled from the inputs by making use of a Normal (defined by the mean and standard deviation) frequency distribution. As soon as a parameter set generates a simulation that satisfies all of the constraints, it is saved to the SPATSIM database (Ndzabandzaba and Hughes, 2017). However, when 1000 output behavioural parameter sets have been found the model terminates. Since the constraints define the uncertainty in the hydrological response behaviour of each of the selected sub-basins (Yadav, et al., 2007; Westerberg et al., 2011, 2014) all of the saved parameter sets represents behavioural responses (Beven, 2012).

4 Results and Discussion

4.1 Uncertainty results

A significant difference between the irrigation coverage is observed for the Mogalakwena catchments giving relatively huge uncertainty. However, the difference between the irrigated areas sourced from the LIMCOM (2013) compared to the areas digitised from Google Earth is marginal, resulting

Table 1. A list of the parameters of the Pitman model including those of the reservoir water balance model (Hughes et al., 2006).

Parameter	Units	Parameter Description
RDF	–	Controls the distribution of total monthly rainfall over four iterations
AI	Fraction	Impervious fraction of sub-basin
PI1 & PI2	mm	Interception storage for two vegetation types
AFOR	%	% area of sub-basin under vegetation type 2
FF	–	Ratio of potential evaporation rate for Veg2 relative to Veg1
PEVAP	mm	Annual sub-basin evaporation
ZMIN	mm month^{-1}	Minimum sub-basin absorption rate
ZAVE	mm month^{-1}	Mean sub-basin absorption rate
ZMAX	mm month^{-1}	Maximum sub-basin absorption rate
ST	mm	Maximum moisture storage capacity
SL	mm	Minimum moisture storage below which no GW recharge occurs
POW	–	Power of the moisture storage – runoff equation
FT	mm month^{-1}	Runoff from moisture storage at full capacity (ST)
GPOW	–	Power of the moisture storage – GW recharge equation
GW	mm month^{-1}	Maximum ground water recharge at full capacity, ST
R	–	Evaporation-moisture storage relationship parameter
TL	months	Lag of surface and soil moisture
CL	months	Channel routing coefficient
DDENS	–	Drainage density
T	m^2 d^{-1}	Ground water transmissivity
S	–	Ground water storativity
GWSlope	–	Initial ground water gradient
AIRR	km^2	Irrigation area
IWR	Fraction	Irrigation water return flow fraction
EffRf	Fraction	Effective rainfall fraction
NIrrDmd	Ml yr^{-1}	Non-irrigation demand from the river
MAXDAM	Ml	Small dam storage capacity
DAREA	%	Percentage of sub-basin above dams
A, B	–	Parameters in non-linear dam area-volume relationship
IrrAreaDmd	km^2	Irrigation area from small dams
CAP	Mm^3	Reservoir capacity
DEAD	%	Dead storage
INIT	%	Initial storage
RES 1–5	%	Reserve supply levels (percentage of full capacity)
ABS	Mm^3	Annual abstraction volume
COMP	Mm^3	Annual compensation flow volume

in small uncertainty. Nevertheless, the uncertainty still had an impact on the low flows of the sub-basins. Shashe catchments BS4 and BS5 have little irrigated areas and the difference between the irrigated areas provided by LIMCOM (2013) and manual digitizing was too small to impact the results.

The overall calculated uncertainty of the Mogalakwena sub-basin varied between 30 and 100 % and between 5 and 100 % for the farm dams and irrigation data respectively. For the Shashe sub-basin, where farm dams and irrigation existed, the calculated uncertainty varied between 5 and 90 % and between 10 and 100 % farm dams and irrigation respectively.

4.2 Water resources simulation of the Mogalakwena sub-basin

Based solely on the uncertainty related to model parameters, the simulated monthly flows ranged between a minimum of 21.53 Mm^3 and a maximum of 24.68 Mm^3; resulting in a mean monthly flow of 22.62 Mm^3. Results indicate that the whole range of flows (i.e. high, medium and low flows) is impacted when uncertainty related to natural parameters is considered. Adding the uncertainty related to water use (i.e. irrigation and farm dams) data resulted in minimum simulated flows of 22.15 Mm^3 and a maximum of 24.99 Mm^3. High flows are more impacted than low flows.

4.3 Water resources simulation for the Shashe sub-basin

The difference between the model parameters and water use uncertainties is barely noticeable for some catchments of the Shashe. This is because the cumulative volumes of the farm dams was too small to have a noticeable impact on the total predictive uncertainty. The same applies to the irrigation water use uncertainty. Considering uncertainty related to the natural parameters of runoff generation, the simulated mean monthly flow was 14.54 Mm^3 with a maximum of 17.15 Mm^3 and a minimum of 11.66 Mm^3. When the uncertainty related to the water use data was added the minimum value of the simulated flow remained the same, whereas the maximum increased to 17.72 Mm^3.

5 Conclusions

The main objective of the study was to demonstrate that estimated water resources are capable of spanning a wide range of plausible or probable values when considering the uncertainties in model parameter estimation, our understanding of the hydrological processes prevalent in a basin and how those processes are represented in the model as well as the use of limited and often poor quality historical observed data to calibrate the model. This study gives insight into how the simulation of the natural water resources of the Mogalakwena and Shashe sub-basins would be expected to vary when expected uncertainty is incorporated into the estimation process. It can be concluded that uncertainty (in this case related to parameter and water use data) plays an important role in the estimation of water resources as demonstrated in the sub-basins. The uncertainty related to the estimation of water use data tends to affect the medium to low flows more than just considering parameter uncertainty. Farm dams absorb the flows during the high flow season, whereas irrigation would be important during the low flow season. This study provides a limited illustration of how the identification and quantification of parameter and water use uncertainties can provide insight into the possible impacts of using a database such as the national water resources assessment study (WR2012) without examining the quality of the data. This study is a step forward in providing insight into water resources estimations based on the data that are routinely used in the assessment processes without examining the possible impact of the uncertainties related to those data.

Data availability. Data not publicly available as they were obtained from a consultant with the express caveat that they would only be used for the project and not shared publicly. However, these data will soon be available on the Limpopo Management Information System (LIMIS) which is managed by the Limpopo River Basin Commission who contracted the consultant to collect the data.

Competing interests. The authors declare that they have no conflict of interest.

Special issue statement. This article is part of the special issue "Understanding spatio-temporal variability of water resources and the implications for IWRM in semi-arid eastern and southern Africa". It is a result of the IAHS Scientific Assembly 2017, Port Elizabeth, South Africa, 10–14 July 2017.

Acknowledgements. This paper forms part of a project titled: "Upstream-Downstream Hydrological Linkages in the Limpopo River Basin" which is funded by the Water Research Commission (WRC) of South Africa (Project No. K5/2439/1) and the CSIR parliamentary grant (ECHS052).

Edited by: Hodson Makurira
Reviewed by: Joel Nobert and Donald Rwasoka

References

Beven, K.: Casual models as multiple working hypotheses about environmental processes, Comptes Rendus – Geoscience, 344, 77–88, 2012.

Busari, O.: Groundwater in the Limpopo Basin: occurrence, use and impact, Environmental Development and Sustainability, 10, 943–957, 2008.

DAFF (Department of Agriculture, Forestry and Fisheries): RSA Irrigation Data, Department of Agriculture, Forestry and Fisheries, Pretoria, South Africa, 2015.

Hughes, D. and Mantel, S. K.: Estimating the uncertainty in simulating the impacts of small farm dams on streamflow regimes in South Africa, Hydrol. Sci. J., 55, 578–592, 2010.

Hughes, D. A.: Southern Africa FRIEND – The Application of Rainfall-Runoff Models in the SADC Region, Water Research Commission Report No. 235/1/97, Pretoria, South Africa, 1997.

Hughes, D. A.: Incorporating ground water recharge and discharge functions into an existing monthly rainfall-runoff model, Hydrol. Sci. J., 49, 297–311, 2004.

Hughes, D. A. and Forsyth, D.: A generic database and spatial interface for the application of hydrological and water resource models, Computers and Geosciences, 32, 1389–1402, 2006.

Hughes, D. A., Andersson, L., Wilk, J., and Savenije, H. H. G.: Regional calibration of the Pitman model for the Okavango River, J. Hydrol. 331, 30–42, 2006.

Hughes, D. A., Kapangaziwiri, E., and Sawunyama, T.: Hydrological model uncertainty assessment in southern Africa, J. Hydrol., 387, 221–232, 2010.

IWR. SPATSIM: Spatial and Time Series Information Modelling Software, Institute for Water Research, Rhodes University, Grahamstown, South Africa, available at: https://www.ru.ac.za/iwr/research/software/, last access: April 2017.

Kapangaziwiri, E., Hughes, D. A., and Wagener, T.: Incorporating uncertainty in hydrological predictions for gauged and ungauged basins in southern Africa, Hydrol. Sci. J., 57, 1000–1019, 2012.

Limpopo Watercourse Commission (LIMCOM): Limpopo River Basin Monograph, Final Monograph. Report No. LRBMS-81137945, 2013.

Matji, M. P. and Görgens, A. H. M.: Update of the Hydrological Model of the Limpopo River Main Stem, Peer-Reviewed Report by the Department of Civil Engineering and Ninham Shand (Pty) Ltd to the Water Research Commission, Pretoria, South Africa, on behalf of the four co-border Countries, 2001.

Ndzabandzaba, C. and Hughes, D. A.: Regional water resources assessments using an uncertain modelling approach: The example of Swaziland, J. Hydrol.: Regional Studies, 10, 47–60, 2017.

Pitman, W. V.: A mathematical model for generating monthly river flows from meteorological data in South Africa, Hydrological Research Unit, Univ. of the Witwatersrand, Report No. 2/73, 1973.

Sawunyama, T., Senzanje, A., and Mhizha, A.: Estimation of reservoir storage capacities in Limpopo River Basin using geographical information systems (GIS) and remotely sensed areas: Case of Mzingwane catchment, Physics and Chemistry of the Earth Parts A/B/C, 31, 935–943, 2006.

Sawunyama, T., Hughes, D. A., and Mallroy, S. J. L.: Evaluation of combined contribution of uncertainty sources to total output uncertainty in water resource estimation in South Africa, IAHS Publication, 347, 2011.

Sayl, K. N., Muhammad, N. S., Yaseen, Z. M., and El-shafie, A.: Estimation the Physical Variables of Rainwater Harvesting System Using Integrated GIS-Based Remote Sensing Approach, Water Resour. Manag., 30, 3299, https://doi.org/10.1007/s11269-016-1350-6, 2016.

Westerberg, I. K., Guerrero, J.-L., Younger, P. M., Beven, K. J., Seibert, J., Halldin, S., Freer, J. E., and Xu, C.-Y.: Calibration of hydrological models using flow-duration curves, Hydrol. Earth Syst. Sci., 15, 2205–2227, https://doi.org/10.5194/hess-15-2205-2011, 2011.

Westerberg, I. K., Gong, L., Beven, K. J., Seibert, J., Semedo, A., Xu, C.-Y., and Halldin, S.: Regional water balance modelling using flow-duration curves with observational uncertainties, Hydrol. Earth Syst. Sci., 18, 2993–3013, https://doi.org/10.5194/hess-18-2993-2014, 2014.

Wilk, J. and Hughes, D. A.: Calibrating a rainfall-runoff model for a catchment with limited data, Hydrol. Sci. J., 41, 3–17, 2002.

Yadav, M., Wagener, T., and Gupta, H. V.: Regionalization of constraints on expected watershed response behaviour, Adv. Water Res., 30, 1756–1774, 2007.

https://doi.org/10.5194/piahs-378-17-2018

Modelling of channel transmission loss processes in semi-arid catchments of southern Africa using the Pitman Model

Vuyelwa Mvandaba[1,2], Denis Hughes[2], Evison Kapangaziwiri[1], Jean-Marc Mwenge Kahinda[1], and Nadia Oosthuizen[1,2]

[1]CSIR, Natural Resources and Environment, P.O. Box 395, Pretoria, 0001, South Africa
[2]Institute for Water Research, Rhodes University, P Bag 94, Grahamstown, 6140, South Africa

Correspondence: Vuyelwa Mvandaba (mvandabav@gmail.com)

Received: 12 November 2017 – Revised: 24 January 2018 – Accepted: 31 January 2018 – Published: 29 May 2018

Abstract. Water availability is one of the major societal issues facing the world. The ability to understand and quantify the impact of key hydrological processes, on the availability of water resources, is therefore integral to ensuring equitable and sustainable resource management. Channel transmission losses are an "under-researched" hydrological process that affects resource availability in many semi-arid regions such as the Limpopo River Basin in southern Africa, where the loss processes amount to approximately 30 % of the water balance. To improve the understanding of these loss processes and test the capability of modelling routines, three approaches using the Pitman model are applied to selected alluvial aquifer environments. The three approaches are an explicit transmission loss function, the use of a wetland function to represent channel-floodplain storage exchanges and the use of a "dummy" reservoir to represent floodplain storage and evapotranspiration losses. Results indicate that all three approaches are able to simulate channel transmission losses with differing impacts on the regional flows. A determination of which method best represents the channel transmission losses process requires further testing in a study area that has reliable observed historical records.

1 Introduction

Channel transmission losses describe surface water and groundwater interactions where streamflow is reduced by evaporation from free water surfaces, evapotranspiration from the recharged river bed and banks, infiltration into channel beds, banks and flood plains and evapotranspiration from riparian vegetation (Smakhtin and Watkins, 1997). Compared to other hydrological processes, information on channel transmission losses remains largely lacking, however the dynamics between alluvial aquifers (alluvium) and channel transmission losses have been covered by several hydrological studies across the world (Boroto and Görgens, 2003; Lange, 2005; Costa et al., 2012; Jarihani et al., 2015). From these studies it is understood that, due to their water storage capabilities, the presence of alluvial aquifers results in significant channel transmission losses in ephemeral river systems found in semi-arid catchments such as the Limpopo River Basin of southern Africa (Boroto and Görgens, 2003). Ultimately, channel transmission loss processes affect the availability of streamflow, hence it is imperative to understand their role in the alluvial aquifers of the Limpopo River Basin.

2 The Pitman Model

The Pitman Model (Hughes, 2004) is a conceptual, monthly time-step and spatially semi-distributed rainfall-runoff simulation model. Its main application, as one of the most extensively used models within southern Africa (Tshimanga et al., 2011), has been as a water resource estimation tool for basin-wide water resource assessments and planning (Kapangaziwiri et al., 2011). Revisions to the model have included incorporation of groundwater recharge and discharge functions as well as the addition of an explicit transmission loss function (Hughes et al., 2013) designed to account for riparian evapotranspiration losses and losses from storage due to sub-

Published by Copernicus Publications on behalf of the International Association of Hydrological Sciences.

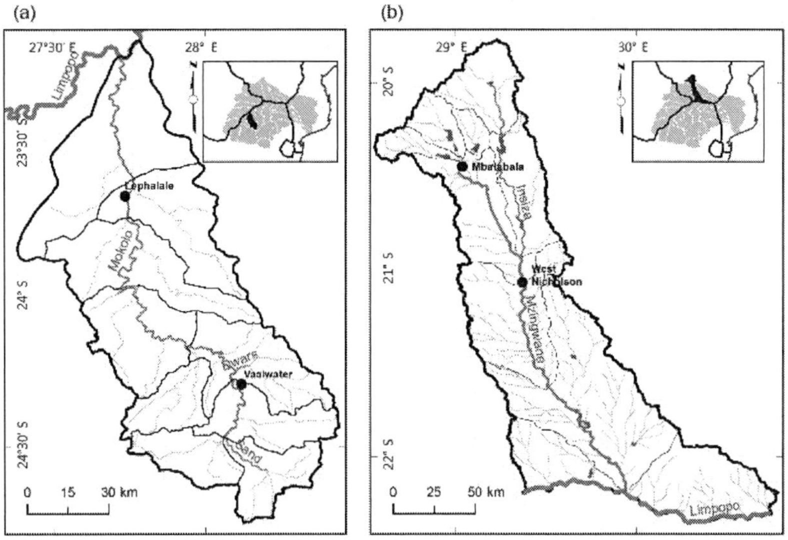

Figure 1. Map of the Mokolo **(a)** and Mzingwane **(b)** sub-basins and their locations in the Limpopo River Basin.

surface outflow to downstream areas. Prior to the development of the transmission loss function, streamflow losses to adjacent alluvial aquifers were simulated using a wetland function to represent channel-floodplain storage exchanges and a "dummy" reservoir to represent floodplain storage and evapotranspiration losses.

2.1 The transmission loss function

The transmission loss function relies on a simple geometry which conceptualises the water balance within the groundwater store. The process of simulating transmission losses in the model assumes that the rate of loss will be due to characteristics of the channel, the head difference between the channel and the groundwater and the transmissivity of the material under the channel (Hughes et al., 2007). The function simulates two groundwater gradients which, when negative, relate the loss of water from channel to aquifer.

2.2 The wetland function

The wetland function is based on a simple water balance approach with water draining into and out of the wetland. The function was added to the model to account for the downstream impact of wetlands and natural lake systems on streamflow (Hughes et al., 2013). Processes associated with wetlands and lakes can exert a considerable influence on downstream flow regimes through attenuation, storage and slow release processes that occur within the water bodies. These processes are critical in understanding the general patterns of runoff generation at the basin scale (Tanner, 2013).

2.3 The reservoir function

The basis of the reservoir function, as a water use component of the Pitman Model, is to simulate the impact of large dams on the basin's streamflow. A study (Hughes et al., 2006) focussing on the impact of climate change and development scenarios on flow patterns in the Okavango River represented channel transmission losses as "dummy" reservoirs based on a quantification of channel lengths and widths of floodplains and swamps – riparian areas that were hypothesised to be fed by seepage from the river, and from which water was assumed to evaporate, were modelled as open water surfaces ("dummy" reservoirs). The study noted that the presence of the "dummy" reservoir reduced streamflow from the Omatako sub-basin to zero flow at the outlet, which is the right behavioural effect however not quite representative as parameter values used were based on inadequate information.

Detailed descriptions of each function are provided by Hughes (2004), Hughes et al. (2007), Kapangaziwiri (2008) and Tanner (2013).

3 Study areas

Two sub-basins of the Limpopo River Basin were selected as study areas, namely the Mokolo (South Africa) and the Mzingwane (Zimbabwe) (Fig. 1).

3.1 The Mokolo sub-basin

The Mokolo sub-basin (8417 km^2) falls within the South African portion of the Limpopo and comprises 9 catchments, namely A42A to A42J. The sub-basin is characterised by an

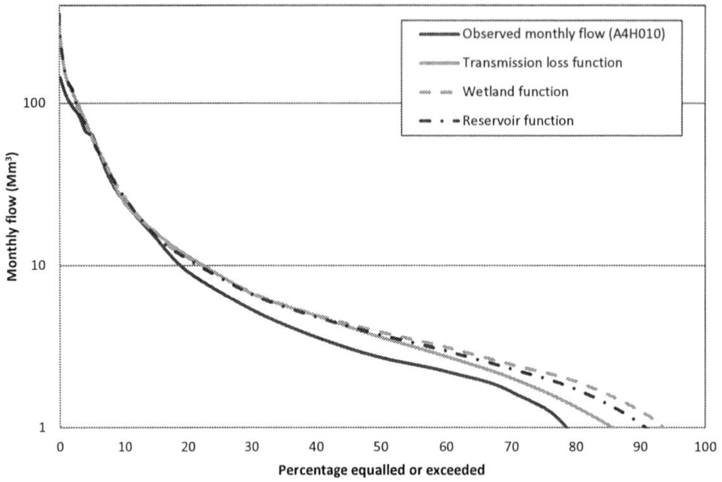

Figure 2. Comparison of flow volumes (Mm^3) at selected flow duration curve percentage points for the Mokolo sub-basin.

arid to semi-arid climate (DWA, 2014) with mean annual precipitation (MAP) and evaporation (MAE) of 350 to 700 and $1900\,\mathrm{mm\,a^{-1}}$, respectively. The flow of the Mokolo River and its tributaries is highly regulated and largely perennial, except for the lower reaches and other lower order streams, which respectively display ephemeral (summer flow) and episodic characteristics (Boroto and Görgens, 1999). The hydrogeology of the Mokolo sub-basin comprises fractured and intergranular secondary aquifers as well as quaternary alluvium which constitute primary alluvial aquifers along the riverbed of the river Mokolo and some of its tributaries (Titus and Rossouw, 2008). Water use in the sub-basin is dominated by agriculture, industrial use, mining and domestic use (DWA, 2012).

3.2 The Mzingwane sub-basin

The Mzingwane sub-basin is situated in the Zimbabwe portion of the Limpopo River Basin (Howard et al., 2013). A semi-arid climate dominates the sub-basin with areas in the upper reaches of the sub-basin receiving a MAP of 300 to $800\,\mathrm{mm\,a^{-1}}$ (Love et al., 2007) and MAE of more than $1800\,\mathrm{mm\,a^{-1}}$ (Howard et al., 2013). The ephemeral Mzingwane River is highly developed. Aquifers in the Mzingwane sub-basin comprise both secondary fractured and primary alluvial aquifers (CSIR, 2003). Major water uses in the sub-basin are for domestic, industrial, mining and agricultural purposes (Mabiza et al., 2007).

4 Model set up and channel transmission loss modelling approach

The Pitman model was set up for both the Mokolo and Mzingwane sub-basins, using the Spatial Time Series Information Modeling (SPATSIM) software (Hughes and Forsyth, 2006). Model inputs for each catchment of the Mokolo sub-basin were based on the regional information available in the Water Resources of South Africa 2005 (WR2005) study (Middleton and Bailey, 2009). Monthly flow volume recorded at gauge A4H010, which is located near the outlet of A42F, was used for validation of the transmission loss simulations given that the flow gauge station at the outlet of the Mokolo River was found to have insufficient data. The data was accessed from the South African Department of Water and Sanitation Hydrological Information System and covers the time period 1979 to 2017. A series of eights nodes, depicting flow gauge stations was used for the setup of the Mzingwane sub-basin. The flow gauges represented include B20, B60, B75, B22, B62, B91, B90 and BUZMZ. Parameter inputs, based on data implemented in the updated Blue Book (Mazvimavi, 2006) and the recent Limpopo River Basin Monograph Study (Howard et al., 2013), were used to populate each of the nodes. Flow gauge B20, which is the most upstream station in the setup, was used for reporting as gauges further downstream were found to be highly impacted by irrigation dams found along the Mzingwane River.

To compare the three transmission loss simulation approaches, each function was ran independently and then the simulation outputs were used to assess the degree of change between the observed flow (no transmission losses included) and a model run with channel transmission losses. Additionally, the outputs simulated by the transmission loss function, wetland function and "dummy" dam function were also compared to determine which function performed better in simulating the estimated channel transmission losses for each sub-basin. The alluvial aquifer storage in each sub-basin and hence the potential quantity of channel transmission losses was calculated by using the dimensions of alluvial aquifers delineated in each sub-basin.

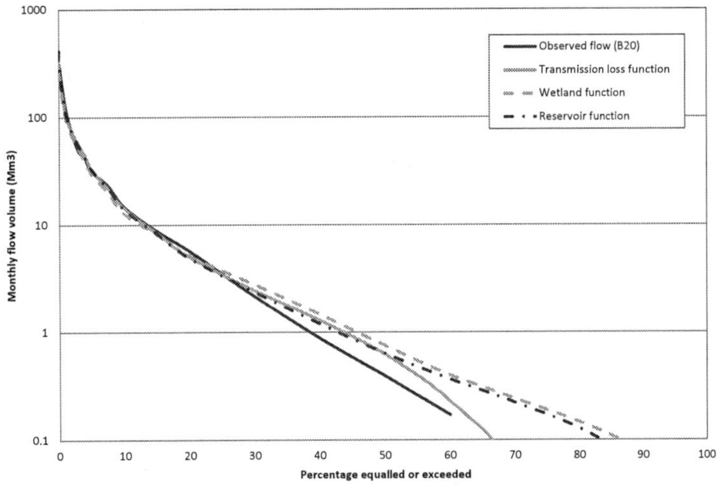

Figure 3. Comparison of observed and simulated flow (Mm^3) for the Mzingwane sub-basin.

5 Results and Discussion

An assessment of the model performance was carried out, after the inclusion of channel transmission losses based on an evaluation of the correlation between observed and simulated flow, based on the recommended statistics for model performance by Moriasi et al., 2007. The model performances for both the Mokolo and Mzingwane sub-basins were satisfactory (CE > 0.50, PBIAS < 25) even before the inclusion of channel transmission losses, however the simulated mean monthly flows were significantly greater than the observed flow. The results of the channel transmission loss simulations are presented on a flow duration curve (FDC) which illustrates the correlation between the simulated monthly flow volume and the observed monthly flow volume.

Channel transmission losses measuring 55 Mm^3 were estimated for catchment A42F, based on results acquired from delineation and characterisation of alluvial aquifers in the Mokolo sub-basin. The transmission loss, wetland and reservoir functions respectively simulated channel losses measuring 2.2 $Mm^3\,month^{-1}$ (26.4 $Mm^3\,a^{-1}$), 56.95 Mm^3 and 54.78 Mm^3. Following the inclusion of transmission losses, simulated MAR measuring 11.71 Mm^3 was respectively reduced to 11.47 Mm^3 and 11.60 Mm^3 by the transmission loss and reservoir functions; MAR increased to 11.72 Mm^3 when losses were added through the wetland function. Observed flow from gauge A4H010 was however over-simulated particularly along the middle and low flows (Fig. 2) because of unaccounted water uses, due to lack of data.

For the Mzingwane sub-basin, channel transmission losses measuring 5 Mm^3 were estimated for node B20. The transmission loss, wetland and reservoir functions respectively simulated channel losses measuring 0.39 $Mm^3\,month^{-1}$ (4.68 $Mm^3\,a^{-1}$), 4.93 Mm^3 and 4.97 Mm^3. Following the inclusion of transmission losses, simulated MAR measuring 6.539 Mm^3 was respectively reduced to 6.486, 6.141 and 6.471 Mm^3 by the transmission loss, wetland and reservoir functions. On the flow duration curve, which is presented on a logarithmic graph (Fig. 3), it is evident that the simulated high flows compare relatively well to the observed high flows however both medium and low flows are over-simulated. Similar to the Mokolo sub-basin, the over-simulation of flow is associated with inadequate representation of water use impacts in the sub-basin. This over-simulation is further evidenced by the relative steepness of the flow duration curve which indicates that high flows are being simulated at a greater frequency than expected.

6 Conclusion

Channel transmission losses play a significant role in the water balance of the Limpopo River Basin, therefore understanding the dynamics which influence loss processes and quantifying the impact on water resources is integral for advancing knowledge and improving water resource management. The aim of simulating channel transmission losses with three different functions of the Pitman Model was to assess the capability of existing loss modelling routines and determine which of the three functions best represents the dynamics of channel transmission losses processes. Transmission loss simulations were conducted in the Mokolo and Mzingwane sub-basins of the Limpopo River Basin and results indicate that all three functions were able to simulate losses, although with differing magnitudes. Observed monthly flow data was used to as a means of validating loss simulations however in both examples, medium and low flows were over-simulated which accounts for water uses that were inefficiently represented due to lack of data. Knowledge of the structure of the transmission loss function dictates that it is better at representing the dynamics of channel transmission losses, as it takes into account the contribution of losses to groundwater recharge whereas the other two functions do not.

Overall, the results from the two modelling examples demonstrate the potential of each approach in reproducing the dynamics of channel transmission losses between channel and alluvial aquifer within an existing sub-basin scale hydrological model. It is believed that better quantification of losses and more efficient qualitative determination of the function which best represents transmission losses, can be attained with more reliable observed data.

Data availability. Data not publicly available as they were obtained from a consultant with the caveat that they would only be used for the project and not shared publicly. However, the data will soon be available on LIMIS which is managed by the Limpopo River Basin Commission who contracted the consultant to collect the data.

Competing interests. The authors declare that they have no conflict of interest.

Special issue statement. This article is part of the special issue "Understanding spatio-temporal variability of water resources and the implications for IWRM in semi-arid eastern and southern Africa". It is a result of the IAHS Scientific Assembly 2017, Port Elizabeth, South Africa, 10–14 July 2017.

Acknowledgements. This paper forms part of a project titled: "Upstream-Downstream Hydrological Linkages in the Limpopo River Basin" which is co-funded by the Water Research Commission, Pretoria, South Africa (Project K5/2439/1) and the CSIR parliamentary grant (ECHS052).

Edited by: Hodson Makurira
Reviewed by: Cosmo Ngongondo and one anonymous referee

References

Boroto, R. A. J. and Görgens, A. H. M.: Hydrological Modelling of the Limpopo River Main Stem, report by: Department of Civil Engineering, University of Stellenbosch and Ninham Shand Consulting Engineers, Department of Water Affairs & Forestry, DWAF Report No. PA000/00/0399, 105 pp., 1999.

Boroto, R. J. and Görgens, A. H. M.: Estimating transmission losses along the Limpopo River: an overview of alternative methods, International Association of Hydrological Sciences Publication, 278, 138–143, 2003.

Costa, A. C., Bronstert, A., and de Araújo, J. C.: A channel transmission losses model for different dryland rivers, Hydrol. Earth Syst. Sci., 16, 1111–1135, https://doi.org/10.5194/hess-16-1111-2012, 2012.

CSIR (Council for Scientific and Industrial Research-Environmentek): Protection and strategic uses of groundwater resources in drought prone areas of the SADC Region: Groundwater situation analysis of the Limpopo River Basin – Final Report, Environmentek Report No. ENV-P-C 2003-026, Pretoria, 2003.

DWA (Department of Water Affairs): Classification of Significant Water Resources in the Mokolo Catchment: Limpopo Water Management Area (WMA) and Crocodile (West) and Marico WMA WP 10506: Information Analysis Report: Mokolo and Matlabas catchments: Limpopo WMA, Report No: RDM/WMA1,3/00/CON/CLA/0112B, Pretoria, South Africa, 2012.

DWA (Department of Water Affairs): Classification of significant water resources in the Crocodile (West), Marico, Mokolo and Matlabas catchment: Management Classes Report, Report No: RDM/WMA1, 3/00/CON/CLA/0612, Pretoria, South Africa, 2014.

Howard, G., Denys, F., Walker, N., and Görgens, A.: Volume C1 – Surface Water Hydrology: Limpopo River BasinMS-81137945, Supplementary Report to Final Limpopo River Basin Monograph, prepared for LIMCOM, 2013.

Hughes, D. A. and Forsyth, D. A.: A generic database and spatial interface for the application of hydrological and water resource models, Comput. Geosci., 32, 1389–1402, 2006.

Hughes, D., Andersson, L., Wilk, J., and Savenije, H. H. G.: Regional calibration of the Pitman model for the Okavango River, J. Hydrol., 331, 30–42, 2006.

Hughes, D., Parsons, R., and Conrad, J.: Quantification of the Groundwater Contribution to Baseflow, WRC Report No. 1498/1/07, Water Research Commission, Pretoria, South Africa, 2007.

Hughes, D. A.: Incorporating ground water recharge and discharge functions into an existing monthly rainfall-runoff model, Hydrolog. Sci. J., 49, 297–311, 2004.

Hughes, D. A., Tshimanga, R. M., Tirivarombo, S., and Tanner, J.: Simulating wetland impacts on stream flow in southern Africa using a monthly hydrological model, Hydrol. Process., 28, 1775–1786, 2013.

Jarihani, A. A., Larsen, J. R., Callow, J. N., McVicar, T. R., and Johansen, K.: Where does all the water go? Partitioning water transmission losses in a data-sparse, multi-channel and low-gradient dryland river system using modelling and remote sensing, J. Hydrol., 529, 1511–1529, 2015.

Kapangaziwiri, E.: Revised parameter estimation methods for the Pitman monthly rainfall-runoff model. MSc Dissertation, Rhodes University, Grahamstown, South Africa, 2008.

Kapangaziwiri, E., Hughes, D. A., Tanner, J. L., and Slaughter, A.: Resolving uncertainties in the source of low flows in South African rivers using conceptual and modelling studies. Conceptual and modelling studies of integrated groundwater, surface water and ecological systems, Proceedings of Symposium H01 held during IUGG 2011, Melbourne, Australia, July 2011, IAHS Publication, 345, 127–132, 2011.

Lange, J.: Dynamics of transmission losses in a large arid stream channel, J. Hydrol., 306, 112–126, 2005.

Love, D., Owen, R., Uhlenbrook, S., van der Zaag, P., and Moyce, W.: The lower Mzingwane alluvial aquifer: managed releases, groundwater-surface water interactions and the challenge of salinity, Interim Paper for WaterNet, 2007.

Mazvimavi, D.: Assessment of Surface Water Resources of Zimbabwe and Guidelines for Planning, Zimbabwe National Water Authority, 2006.

Mabiza, C. C., Van der Zaag, P., Manzungu, E., and Ahlers, R.: The making of a catchment plan: experiences of the Mzingwane Catchemnt, Zimbabwe, available at: http://hdl.handle.net/10568/21290 (last access: 1 April 2017), 2007.

Middleton, B. J. and Bailey, A. K.: Water Resources of South Africa, 2005 study, Water Research Commission Report Number TT 380/08, Water Research Commission, Pretoria, South Africa, 2009.

Smakhtin, V. Y. and Watkins, D. A.: Low-flow Estimation in South Africa, WRC Report 494/1/97, Pretoria, South Africa, 1997.

Tanner, J. L.: Understanding and Modelling of Surface and Groundwater Interactions, PhD Dissertation, Rhodes University, Grahamstown, South Africa, 2013.

Titus, R. and Rossouw, T.: Groundwater Reserve Determination-Study for Mokolo (A42) Catchment, Water Geosciences Consulting, Pretoria, South Africa, 2008.

Tshimanga, R. M., Hughes, D. A., and Kapangaziwiri, E.: Initial calibration of a semi-distributed rainfall runoff model for the Congo River basin, Phys. Chem. Earth, 36, 761–774, 2011.

https://doi.org/10.5194/piahs-378-23-2018

Assessment of impacts of climate change on surface water availability using coupled SWAT and WEAP models: case of upper Pangani River Basin, Tanzania

Peter Kishiwa, Joel Nobert, Victor Kongo, and Preksedis Ndomba

Department of Water Resources Engineering, University of Dar es Salaam, Dar es Salaam, Tanzania

Correspondence: Joel Nobert (njoelk@yahoo.com)

Received: 11 November 2017 – Revised: 12 March 2018 – Accepted: 19 April 2018 – Published: 29 May 2018

Abstract. This study was designed to investigate the dynamics of current and future surface water availability for different water users in the upper Pangani River Basin under changing climate. A multi-tier modeling technique was used in the study, by coupling the Soil and Water Assessment Tool (SWAT) and Water Evaluation And Planning (WEAP) models, to simulate streamflows under climate change and assess scenarios of future water availability to different socio-economic activities by year 2060. Six common Global Circulation Models (GCMs) from WCRP-CMIP3 with emissions Scenario A2 were selected. These are HadCM3, HadGEM1, ECHAM5, MIROC3.2MED, GFDLCM2.1 and CSIROMK3. They were downscaled by using LARS-WG to station scale. The SWAT model was calibrated with observed data and utilized the LARS-WG outputs to generate future streamflows before being used as input to WEAP model to assess future water availability to different socio-economic activities. GCMs results show future rainfall increase in upper Pangani River Basin between 16–18 % in 2050s relative to 1980–1999 periods. Temperature is projected to increase by an average of 2 °C in 2050s, relative to baseline period. Long-term mean streamflows is expected to increase by approximately 10 %. However, future peak flows are estimated to be lower than the prevailing average peak flows. Nevertheless, the overall annual water demand in Pangani basin will increase from 1879.73 Mm^3 at present (2011) to 3249.69 Mm^3 in the future (2060s), resulting to unmet demand of 1673.8 Mm^3 (51.5 %). The impact of future shortage will be more severe in irrigation where 71.12 % of its future demand will be unmet. Future water demands of Hydropower and Livestock will be unmet by 27.47 and 1.41 % respectively. However, future domestic water use will have no shortage. This calls for planning of current and future surface water use in the upper Pangani River Basin.

1 Introduction

Surface water is a natural resource occurring on the earth's surface and plays an essential role in sustaining mankind and other forms of life (Babel et al., 2005). Decisions regarding water allocations are increasingly important as river basins are further developed and managed to meet social, economic, and environmental needs (Grabow and McCornick, 2007). The ever-increasing anthropogenic activities as a result of growing population are putting immense pressure on available surface water resources (Gao et al., 2017). Climate change presents a significant additional challenge to the achievement of sustainable water management around the globe (Ostefeld et al., 2012). Climate change will likely affect the surface water resources due to the expected changes in precipitation and evapotranspiration and the spatio-temporal distribution of these essential water balance components (Kumar et al., 2017).

A rise in air temperature leads to higher evapotranspiration rates of which further increase the demand for irrigation water (Wang et al., 2012). Such change in climate influences the partitioning of rainfall leading to alteration of water balance. Streamflows in upper Pangani River Basin has continued to decrease from several hundreds to 43.37 $m^3\,s^{-1}$ as received at the outlet of catchment (PBWO/IUCN, 2009) i.e. the Nyumba ya Mungu Reservoir (cf. Fig. 1). This decreas-

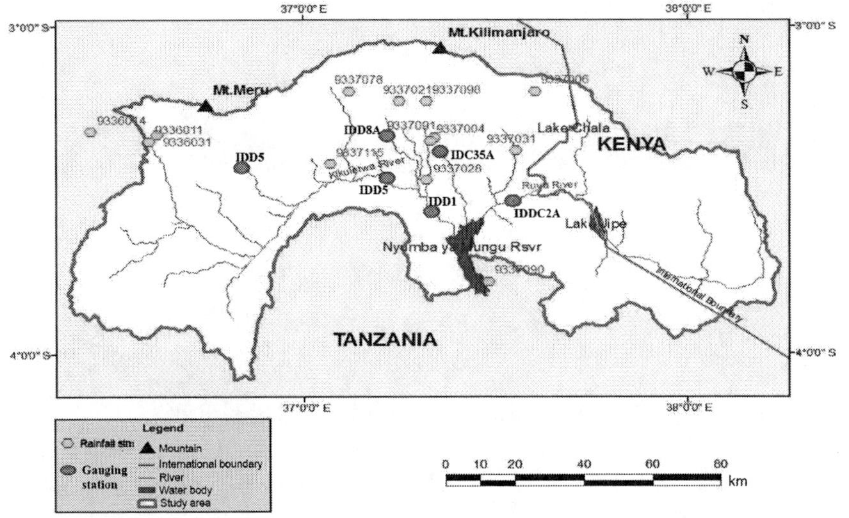

Figure 1. Location of Ruvu and Kikuletwa catchments (upper Pangani).

ing trend is predicted to continue if no proper measures are taken (Funk et al., 2005).

Previous efforts at assessing impacts of climate change on water resources in the upper Pangani Basin has often focused mainly on using outputs from General Circulation Models (GCMs) as input to hydrological models to assess impact of climate change on water resources. In this study, we have taken a step further by using a model-coupling approach through taking advantage of SWAT's capability in representing hydrological processes and WEAP's strength in water allocation using scenarios. Hydrological models, like the SWAT model, are designed to provide an understanding of response of catchments to hydrological events while water resource planning models e.g WEAP are primarily customized for water allocation within a water management context e.g., supply and demand decisions (Yates et al., 2009). Thus, the coupling of SWAT and WEAP models provides a useful platform for analyzing impacts of climate change on water resources – from both management and planning perspectives. The output of GCM provides insights on future climate but at a coarser resolution. Such coarse spatial scales of GCM's are not useful at catchment and or river basin scales of the order of upper Pangani River Basin, with an area of 12 829 km^2. This requires downscaling outputs of GCM's to desired scales of which the Long Ashton Research Station Weather Generator (LARS-WG) approach was used (Semenov et al., 1998) in this study.

The study highlights the potential dynamics of managing water resources in the upper Pangani River Basin in the context of climate change. It also contributes to knowledge and literature on enhancing our understanding of climate change and its potential impacts on surface water in the Basin.

2 Materials and methods

2.1 The study area and data

Kikuletwa and Ruvu catchments (upper Pangani) occupy the area upstream of Nyumba ya Mungu (NyM) reservoir in the Northern part of Pangani River Basin as indicated in Fig. 1. They are located in North-Eastern part of Tanzania and southern part of Kenya, between latitudes and longitudes of 3°00′/4°10′ S and 36°38′/38°20′ E respectively. The two catchments form the headwater of the Pangani River. Their drainage starts from the slopes of Mountains of Kilimanjaro and Meru. They drain water from many small streams which eventually forms the two main rivers of Kikuletwa and Ruvu as indicated in Fig. 1. These two rivers finally discharge water into Nyumba ya Mungu reservoir.

The data used for this study includes daily discharge, rainfall data, temperature and climatic data. A total of 13 rainfall stations were used for the analysis. The criteria used for the selection of the rainfall stations were good spatial distribution, small percentage of missing data and concurrent data. Record length for the data varied from station to station, the concurrent period being 1972–1990. The location of rainfall and river gauging stations used in this study is as indicated in Fig. 1.

2.2 Climate change scenarios and downscaling

Six common GCMs; HadCM3, HadGEM1, ECHAM5, MIROC3.2MED, GFDLCM2.1 and CSIROMK3 from WCRP-CMIP3 with SRES emission scenario A2 were selected. The GCMs were downscaled using LARS-WG to station scale. In order to run the LARS-WG model and downscale GCM data for future periods, two files were created,

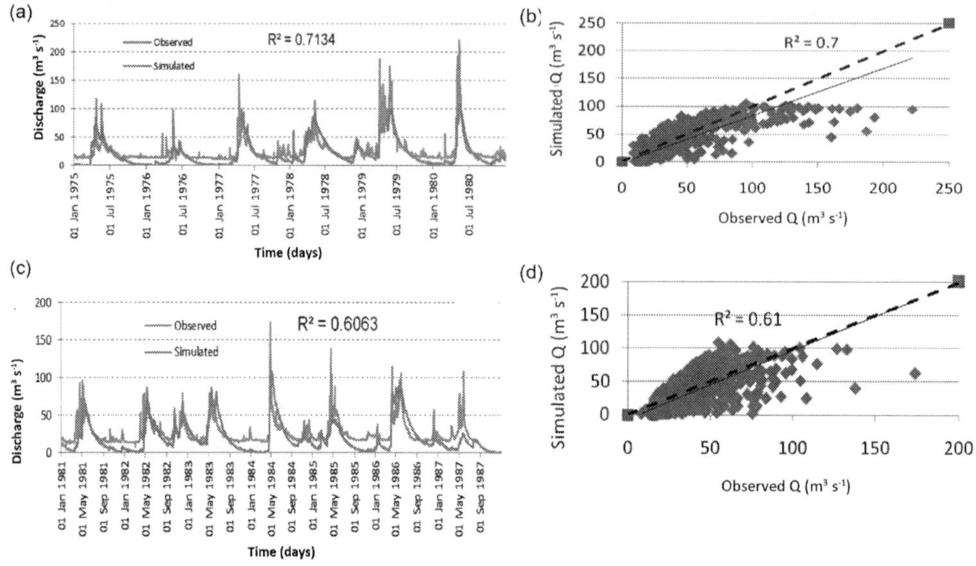

Figure 2. Calibration **(a–b)** and validation **(c–d)** results.

one file defines the behavior of the climate in the past and the other is climate change scenario file. The base period of 50 years (1961–2010) was selected to establish the baseline time series of climate data and then calibrated for the common period of 1975–1990. The performance of LARS-WG was validated with observed records from 1981 to 1990 at each weather station. There are a number of systematic biases from model to model in simulating present day climate and therefore, in this study the average of output from the six selected GCMs was used for the analysis.

2.3 Setting hydrological model

Soil and Water Assessment Tool (SWAT) model was used for hydrological modelling. The model was set-up, calibrated and validated before being used for gap filling extending the flow records and simulating impact of climate change. The 1972 to 1974 period was used as warm-up period, the 1975 to 1980 period for calibration and the 1981 to 1987 period for validation. The sensitivity analysis resulted in 12 sensitive SWAT parameters which were then optimized. SWAT's performance assessment was based on guidelines proposed by Moriasi et al. (2007).

2.4 Water demand and allocation analysis

The main competing water uses in the upper Pangani catchment are domestic, irrigation, livestock, and hydropower. The current and future water demands were estimated based on the existing water permits from the water board, population data and potential irrigable areas. WEAP model was used for water allocation modelling based on selected potential future water use scenarios including population growth, expansion of irrigation, maximizing hydro-electric power (HEP), livestock population growth and climate change.

3 Results and discussions

3.1 SWAT model calibration and validation

Discharge data at gauging station 1DD1 in Kikuletwa River (cf. Fig. 1), for the period 1975 to 1980, was used to calibrate the SWAT model while validation was from 1981 to 1987 as indicated in Fig. 2. The SWAT model performed comparatively well during calibration and validation, with an R^2 of 0.71 and 0.60 respectively. Also Nash-Sutcliffe (NSE), percent bias (PBIAS), and ratio of the root mean square error to the standard deviation of the measured data (RSR) were 0.62, 22.4 and 0.62, respectively, for the calibrated model. The model did not mimic very well both low and high flows as indicated in both line and scatter plots. The inability of the model to capture baseflows could be as a result of an inherent characteristic of the catchment which is typified by presence of springs which eventually maintain a relatively high level of base flows throughout the year.

3.2 Current and future water availability

The mean outputs of the GCM's indicate that upper Pangani catchment will experience increased rainfall as indicated in Fig. 3. The increase is expected to range from 16 to 18 % and temperature is projected to increase by 2° C in 2060s relative to the base year 2011. Based on the modelling results, the long-term mean streamflows are expected to increase by 10 %.

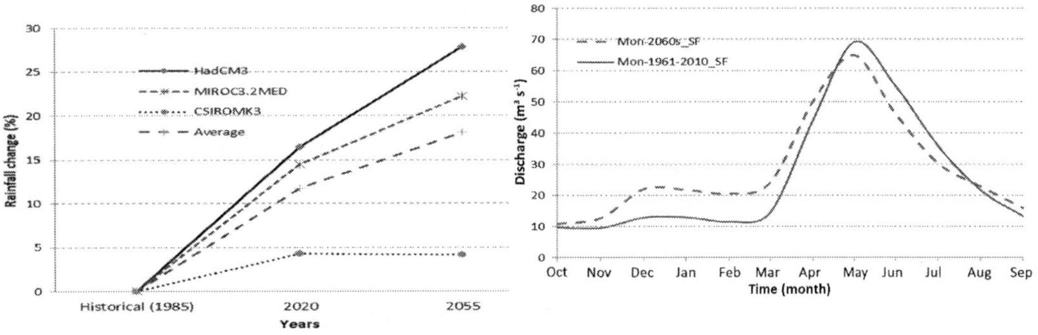

Figure 3. GCM's delta change and future mean monthly stream flows at IDDC1 in Kikuletwa.

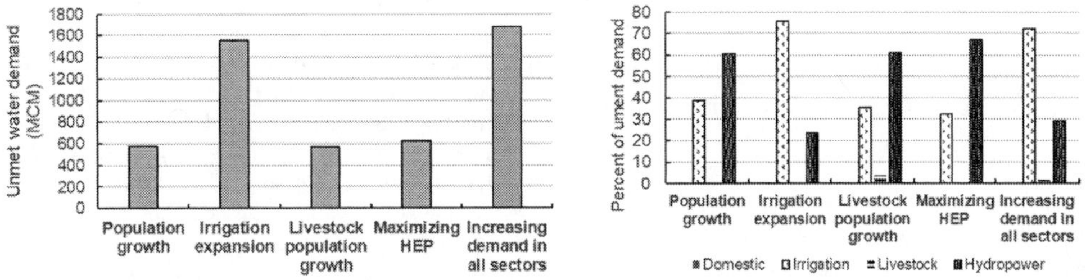

Figure 4. Annual unmet total water demand and corresponding percentages for different sectors.

3.3 Water demand and allocation under selected scenarios

Scenario 1: Impact of population growth. Under this scenario population growth rate used varied from region to region in the range of 1.8–3.2 %. All other sectors were assumed to grow at insignificant rates and therefore their demand for 2060s will be almost equal to the current demand. The sectors that will be highly affected by the impact of population growth will be HEP and Irrigation. HEP demand will be unmet by 350.83 Mm^3 (60.6 %) of the future unmet demand. Irrigation demand will be unmet by 224.37 Mm^3 (38.75 %) of the future unmet demand. However, the domestic water demand will be fully met under the scenario.

Scenario 2: Impact of irrigation expansion. Under the scenario, the area under irrigation is assumed to expand from the current area under irrigation of 49 143 to 79 849 ha which is the potential irrigable area. The irrigation sector will suffer most under the scenario. It is expected to have water shortage of 1177.01 Mm^3 (75.90 %) of the total unmet demand. The other demand which is expected to have significant water shortage under this scenario is HEP with 23.72 % unmet demand (Fig. 3.4). The domestic sector will not be affected by this scenario but livestock sector will experience a little impact of 0.38 % unmet demand.

Scenario 3: Impact of livestock population growth. In this scenario livestock population growth rate of 2.6 % was used to estimate the future population and water demand. It was noted that this scenario has the least impact on the water demand as compared to all other scenarios. Livestock population growth scenario is expected to cause an annual water shortage of 568.43 Mm^3. Much of this shortage will be on the HEP demand which will experience a shortage of 345.79 Mm^3 (60.83 %) of the total unmet demand under this scenario (Fig. 4).

Scenario 4: Maximizing HEP production. The scenario describes what could be expected if hydropower generation was prioritized to operate at maximum capacity, current turbines discharge is 35 $m^3\,s^{-1}$ and the maximum capacity is 42 $m^3\,s^{-1}$. This will result to unmet water demand in the catchment of 623.49 Mm^3 (29.68 %) of the total demand (Fig. 4). HEP will be most affected with unmet demand of 419.17 Mm^3 (67.23 % of total deficit). Irrigation will experience unmet demand of 32.2 % followed by livestock sector (0.6 %).

Scenario 5: Increasing demand for all sectors. The unmet total future demand due to increasing demand of all sectors is estimated to be about 1674 Mm^3. The impact of all future demands expansion and climate change is expected to be severe in the irrigation sector with unmet demand of 1206.66 Mm^3 which is 72.09 % of total unmet demand. The other sectors which will be affected by the scenario are HEP (26.4 %) and livestock (1.5 % unmet). Domestic water demand will be fully met under the scenario.

4 Conclusions

Model results indicate an increase in future rainfall in the range between 16 to 18 % while temperature is projected to increase by 2 °C in 2060s relative to the base year 2011. Long-term mean streamflows is expected to increase by approximately 10 %. However, future peak flows are estimated to be lower than the prevailing average peak flows, while base flows in the dry months between November and February are expected to increase. Nevertheless, the overall annual water demand in Pangani basin will increase from 1879.73 Mm^3 at present (2011) to 3249.69 Mm^3 in the future (2060s). Therefore, the aggregated effect of climate change together with increase in future water demand will result in an annual unmet demand of 1673.80 Mm^3 (51.51 % of future demand) and mostly affecting the Irrigation sector (72.09 % of total unmet demand). This will increase the pressure on water demand due to increased unmet demands especially for irrigation.

Data availability. Data is not publically available at the moment. It was requested for this research purpose under the condition that the raw data should not be shared publically. The custodian of the raw data is the Tanzania Meteorological Agency and the Pangani basin water board.

Competing interests. The authors declare that they have no conflict of interest.

Special issue statement. This article is part of the special issue "Understanding spatio-temporal variability of water resources and the implications for IWRM in semi-arid eastern and southern Africa". It is a result of the IAHS Scientific Assembly 2017, Port Elizabeth, South Africa, 10–14 July 2017.

Acknowledgements. The authors would like to acknowledge the financial support from Climate Change Impacts on Ecosystem Services and Food Security in Eastern Africa (CHIESA).

Edited by: Webster Gumindog
Reviewed by: Timothy Dube and two anonymous referees

References

Babel, M. S, Gupta, A. D, and Nayak, D. K.: A Model for Optimal Allocation of Water to Competing Demands, Water Resour. Manag., 19, 693–712 2005.

Funk, C., Senay, G., Asfaw, A., Verdin, J., Rowland, J., Michaelson, J., Eilerts, G., Korecha, D., and Choularton, R.: Recent Drought Tendencies in Ethiopia and Equatorial-Subtropical Eastern Africa, Washington DC, FEWS-NET, 2005.

Gao, J., Li, F., Gao, H., Zhou, C., and Zhang, X.: The impact of land-use change on water-related ecosystem services: a study of the Guishui River Basin, Beijing, China, J. Clean. Prod., 163, 148–155, 2017.

Grabow, G. L. and McCornick, P. G.: Planning for Water Allocation and Water Quality Using a Spreadsheet-Based Model, J. Water Res. Plan. Man., 133, 560–564, 2007.

Kumar, N., Tischbeinb, B, Kuschec, J., Laux, P., Bege, M. K., and Bogardi, J. J.: Impact of climate change on water resources of upper Kharun catchment in Chhattisgarh, India, Journal of Hydrology: Regional Studies, 13, 189–207, 2017.

Moriasi, D. N., Arnold, J. G., Van Liew, M. W., Bingner, R. L., Harmel, R. D., and Veith, T. L.: Model evaluation guidelines for systematic quantification of accuracy in watershed simulations, T. ASABE, 50, 885–900, 2007.

Ostfeld, A., Barchiesi, S., Bonte, M., Collier, C., Cross, K., Darch, G., Farrel, T., Smith, M., Vicory, A., Weyand, M., and Wright, J.: Climate impacts on river basin and freshwater ecosystems: some observations on challenges and emerging solutions, J. Water Clim. Change, J. Water Clim. Change, 3, 171–184, https://doi.org/10.2166/wcc.2012.006, 2012.

PBWO/IUCN: The Pangani River Basin: A Situational Analysis, 2nd Edition, 12, 82 pp., 2009.

Semenov, M. A., Brooks R. J., Barrow, E. M., and Richardson, C. W.: Comparison of WGEN and LARS-WG Stochastic Weather Generators for Diverse Climates, Climate Research, 10, 95–107, 1998.

Wang, K., Dickinson, R. E., and Liang, S.: Global atmospheric evaporative demand over land from 1973 to 2008, J. Climatol., 25, 8353–8361, https://doi.org/10.1175/JCLI-D-11-00492.1, 2012.

Yates, D., Purkey, D., Sieber, J., Huber-Lee., A, Galbraith, H., West, J., Herrod-Julius, S., Young, C., Joyce, B., and Rayej, M.: Climate Driven Water Resources Model of the Sacramento Basin, California, J. Water Res. Plan. Man., 135, 303–313, 2009.

https://doi.org/10.5194/piahs-378-29-2018

Open Access
PIAHS

Spatio-temporal variations of hydrochemical and isotopic patterns of groundwater in hand-dug wells: the Cuvelai-Etosha Basin, Namibia

Josefina Tulimevava Hamutoko[1], **Heike Wanke**[1], **Matthias Beyer**[2], **Marcel Gaj**[2,3], **and Paul Koeniger**[2]

[1]Geology Department, University of Namibia (UNAM), 340 Mandume Ndemufayo Avenue, Windhoek, Namibia

[2]Federal Institute for Geosciences and Natural Resources (BGR), Hannover, Germany

[3]Global Institute for Water Security, School of Environment and Sustainability, University of Saskatchewan, Saskatoon, Canada

Correspondence: Josefina Tulimevava Hamutoko (jhamutoko@gmail.com)

Received: 10 November 2017 – Revised: 15 March 2018 – Accepted: 17 March 2018 – Published: 29 May 2018

Abstract. The rural population in parts of the Cuvelai-Etosha Basin (CEB) in Namibia depends on groundwater as a source for drinking and livestock watering. The aim of this study is to investigate spatial-temporal patterns for understanding water origins and recharge processes of perched aquifers in the CEB. The study uses hydrochemical data and water stable isotope signatures (^{18}O and ^{2}H) of samples collected during (10) ten field campaigns over a three-year period (from 2014 to 2016) originating from two regions within the basin (Ohangwena and Omusati). A clear distinction between the shallow groundwater in the Omusati and Ohangwena regions documented by TDS values, hydrochemical water types as well as the isotopic compositions of water. These differences are influenced mainly by their location in the landscape, depth to water table, the type of well infrastructure and aquifer material. Spatial and temporal variations indicate that even though these perched aquifers are both within the same basin, they are chemically different because of the rock materials in which their flow. Therefore, these differences in the water hydrochemical composition as well as the processes governing perched aquifers must be taken into account when planning groundwater management in the basin.

1 Introduction

More than half of the rural communities in the Namibian part of the Cuvelai-Etosha Basin (CEB) depend on groundwater as a main source of domestic water supply. The majority of people live in rural areas and are supplied by groundwater wells that tap both shallow and deep aquifers. The shallow groundwater which is trapped on both local, discontinuous impermeable layers (perched aquifers) and shallow regional aquifers is commonly tapped by traditional hand-dugs wells that are more affordable and cheap in comparison with deep boreholes. However, because of the shallow depth to water table and assumed small catchment of these local aquifers, the perched aquifers face both naturally and human induced challenges, e.g., population growth and agricultural activities; lack of proper water abstraction infrastructure; lack of improved sanitation facilities and changes in the climatic regime that may promote both contamination and over-exploitation of the water resources of the shallow aquifers (Hamutoko et al., 2016). Different spatial and temporal scales are particularly important for arid environments due to the high heterogeneity associated with these environments. Ledesma-Ruiz et al. (2015) highlighted that spatial and temporal variability evaluation of major ions are commonly used in groundwater to give insight to aquifer's heterogeneity and processes controlling the groundwater chemistry. Wanke et al. (2014) described water quality from hand dug wells of different areas in Namibia, and concluded that the water in hand-dug wells is not acceptable for drinking and domestic purposes mainly because of its high turbidity, total dissolved solids (TDS), fluoride and nitrate values. Hamutoko et al. (2017) characterized the hydrochemical and

Published by Copernicus Publications on behalf of the International Association of Hydrological Sciences.

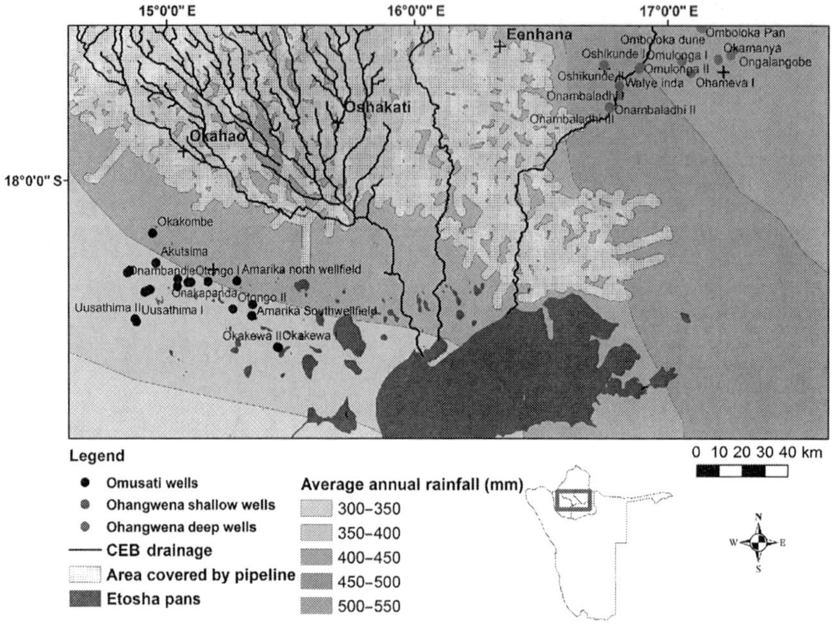

Figure 1. Location of the study area and sampling sites. The map also shows area that is connected to the national water supply pipeline.

isotopic composition of the hand-dug wells in the Ohangwena region and concluded that the groundwater is bicarbonate dominated while the isotope signatures imply recharge through both fast infiltration and diffuse recharge. Nevertheless, spatio-temporal variations of both hydrochemical composition and isotopic signatures are still not well understood and this hinders an effective management of these important shallow water resources. Hence, in this the study we sought to understand the spatial and temporal variations of major ions and water stable isotope signatures (^{18}O and ^{2}H), to determine the predominant processes that contribute to groundwater recharge and evolution of hydrochemistry of hand-dug wells in the CEB.

2 Study area

The study was carried out in the as part of the CEB (Fig. 1). CEB is the Namibian part of the Cuvelai catchment which is a huge sedimentary basin extending from the southern Angolan highlands into north central Namibia covering an area of about 165 000 km^2 (Lindenmaier et al., 2014). Two core study areas were defined, namely Ohangwena and Omusati regions (Table 1). The drainage system is directed towards the saline Etosha pan which is the lowest point in the basin (Fig. 1).

The Ohangwena multi-layered aquifers (KOH) has three aquifers. In this area KOH consist of local perched aquifers (KOH-0) that lie between 0 and 40 m on top, underlain by the regional Ohangwena 1 aquifer (KOH-I) is between 60 and 160 m below groundwater level (m b.g.l.) and at the bottom is the Ohangwena 2 aquifer (KOH-II) which is between 250 to 350 m b.g.l. (Bittner, 2006). The Omusati multi-zoned aquifer (KOM) has a shallow depth to water table, ranging between 10 to 100 m with perched aquifers occurring in a fewer places as compared to the KOH region. The Ohangwena region is characterized by two types of hand-dug wells: shallow wells (funnel-shaped hand-dug wells) locally known as "Omifima" and the deep wells (cylindrical-shaped hand-dug wells) known as "Eendungu". In the Omusati region all hand-dug wells are of the funnel-shaped type. In the Ohangwena region all hand-dug wells are tapping the perched aquifers while in the Omusati region some hand-dug wells are tapping both perched aquifers and the shallow parts of the regional aquifer KOM.

3 Sampling Methods and Laboratory Analysis

Water samples were collected from 48 hand-dug wells, from the two core study regions. Sampling took place over a period of three years (2014–2016) within ten campaigns. Some wells were not sampled at all times because they were not in use or/and dry during the particular field campaign; thus sampling would have been unrepresentative. In the field the physical parameters pH, electrical conductivity (EC), redox potential (Redox), oxygen content (O_2) and temperature (T) were measured with Hach field portable instruments (HQ 11d pH meter, HQ 14d conductivity meter, HQ 40d multimeter). The water level was measured with an electrical contact gauge. Major ions and stable isotopes samples were collected following standard procedures. Chemical analy-

Table 1. Comparison of the two core sites and main results. Source for climate data Mendelsohn et al. (2013).

	Omusati wells	Ohangwena shallow wells	Ohangwena deep wells
Precipitation (mm a^{-1})	350–400	450–500	500–550
Potential evapotranspiration (mm a^{-1})	2600–3000	2400–2600	2400–2600
Local geology	Sand, clay and calcrete/dolocrete and large evaporitic deposits	Fine sands, clay and calcrete	Medium to coarse grained Sand
Water level (m b.g.l.)	0–10	0–5	5–30
Main water type	$CaSO_4$, $NaSO_4$/NaCl, $CaMgHCO_3$	Mixed Cation-HCO_3 NaK-mixed Anions	$CaMgHCO_3$, $NaKHCO_3$
Isotopic composition (‰)	$\delta^{18}O$: −7.81 to 1.32 $\delta^{2}H$: −54.77 to −13.28	$\delta^{18}O$: −6.95 to 0.86 $\delta^{2}H$: −49.14 to −7.64	$\delta^{18}O$: −9.00 to −5.82 $\delta^{2}H$: −61.51 to −37.18

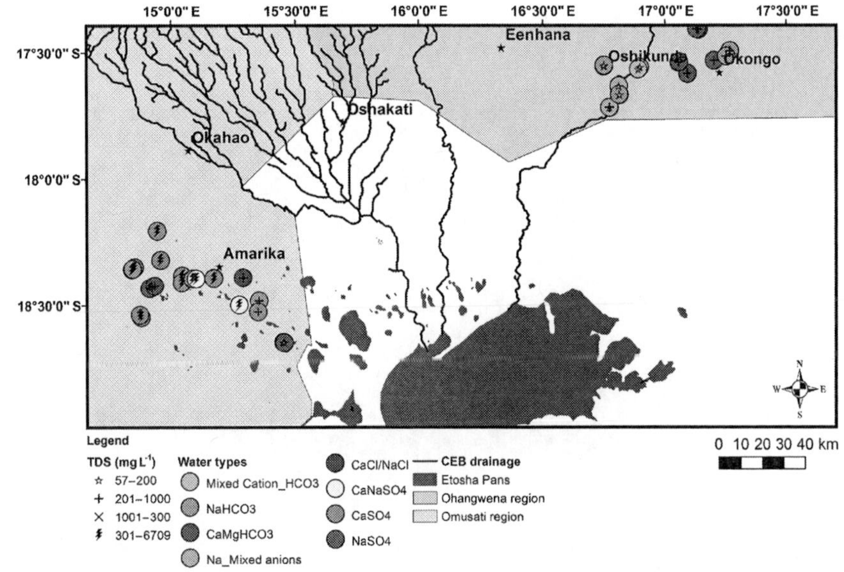

Figure 2. Spatial distribution of TDS and water types in the study area.

ses were performed at the Analytical Laboratory Services in Windhoek, Namibia and at the hydrochemistry laboratory of German Federal Institute for Geosciences and Natural Resources (BGR) in Hanover, Germany. The methods used at each laboratory is discussed in details by Hamutoko et al. (2017). Stable isotopes were measured at the University of Namibia (UNAM) and BGR laboratories using an off-axis integrated cavity output spectroscope (OA-ICOS, Los Gatos DLT-100) and a cavity ring down spectroscope (CRDS, Picarro L2120-i) respectively.

4 Results

4.1 Spatial variations in water chemistry and isotopes

Hydrochemical and isotopic data reflect spatial variability between samples from Omusati and Ohangwena regions. Average values for each elements (major ions and isotopes) were calculated for each sampled site and the average values were used to show spatial variability. Relatively high TDS values are observed in Omusati compared to the Ohangwena region (Fig. 2). In Ohangwena bicarbonate is the dominant anion whereas in Omusati the groundwater is dominated by sulphate, whereas cation varies between calcium and magne-

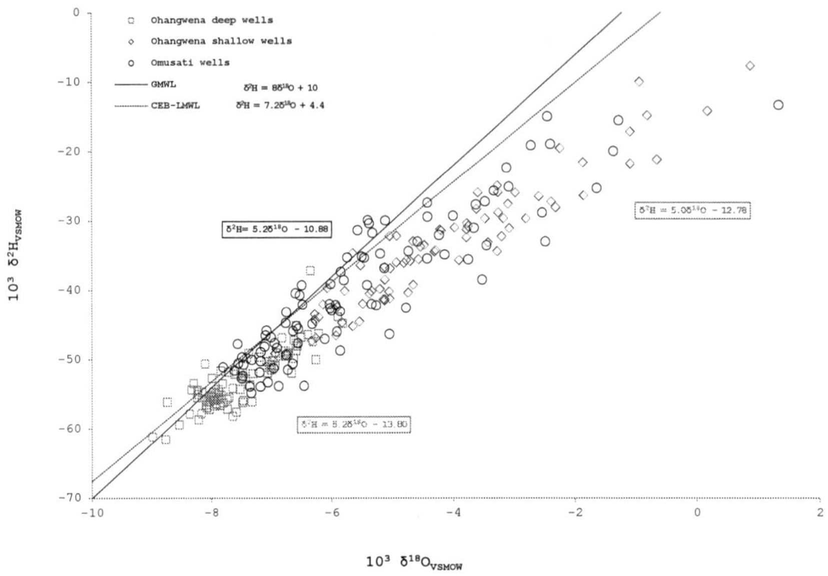

Figure 3. δ^2H vs. $\delta^{18}O$ plot of all collected samples including the global meteoric (GMWL) and local meteoric (LMWL) water lines.

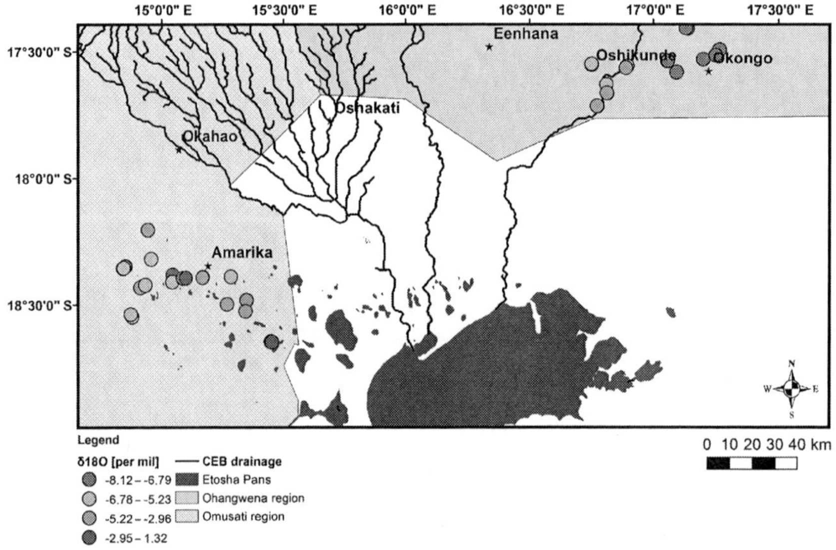

Figure 4. Spatial distribution of $\delta^{18}O$ [per mil] in the Ohangwena and Omusati regions.

sium (Ca-Mg) dominated to sodium and potassium (Na-K) dominated water (Fig. 2). The stable isotope values of all collected samples are plotting close to the Global Meteoric Water Line (GMWL) (Fig. 3). The samples from the Omusati region cover a wide isotopic range from depleted to highly enriched values while in Ohangwena there are two distinct groups. One group represents samples from deep wells that are depleted in $\delta^{18}O$ and δ^2H values, and the second group comprises samples from shallow wells which are enriched in $\delta^{18}O$ and δ^2H values. Spatial distribution of $\delta^{18}O$ shows that the deep wells are more depleted than the open shallow wells (Fig. 4).

In Ohangwena the deep wells plot of on regression line with equation $\delta^2H = 5.2\ \delta^{18}O - 13.80$ while shallow wells plot of regression line with equation $\delta^2H = 5.0\ \delta^{18}O - 12.78$. The samples from Omusati plot on regression line with equation $\delta^2H = 5.2\ \delta^{18}O - 10.88$. These lines intersect the LMWL at (−8.7, −58.2), (−7.8, −51.8) and (−8.6, −57.5) for Ohangwena deep wells, Ohangwena shallow wells and Omusati wells respectively.

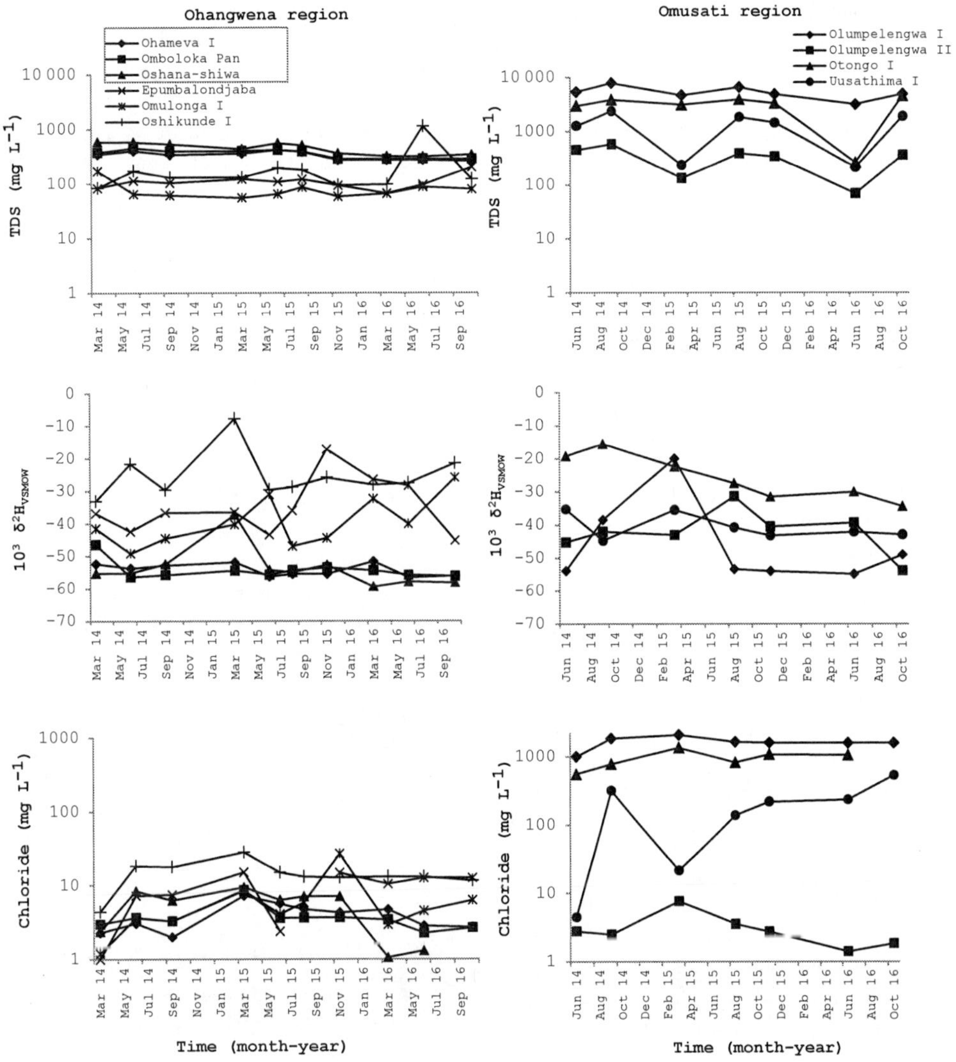

Figure 5. Temporal variation of TDS, δ^2H and chloride concentrations of shallow groundwater in Ohangwena and Omusati regions. The villages in the red box represent deep wells from the pan/depression.

4.2 Temporal variations in Cl, TDS and δ^2H

Temporal patterns were observed for Cl, TDS and δ^2H for ten hand-dug wells. In Ohangwena region, the change in isotope values and chloride concentration follows a similar trend, during rainy season (e.g. March to June in 2015 and 2016) chloride decline and isotope values become depleted while during dry season, isotopes become enriched and chloride concentration increases (Fig. 5). In the Omusati region TDS increases during dry seasons and reduces during rainy season. A change in Cl content is observed between the different years of sampling, however no trend could be identified.

5 Discussions

This study reveals a clear distinction between the shallow groundwater in the two regions Omusati and Ohangwena. This is documented by the TDS values, hydrochemical water types as well as the isotopic compositions of water. TDS is generally higher in Omusati than in Ohangwena. The spatial heterogeneity is controlled mainly by their location in the landscape, depth to water table, the type of well infrastructure, lithological and climatic factors. The main water types ($CaSO_4$, $NaSO_4$, and $CaNaSO_4$) in Omusati region typically originate from dissolution of evaporitic rocks (i.e. gypsum) and evaporation and they are associated with discharge areas. Thus pans in Omusati are characterized as discharge zones rather than recharge zone. Furthermore, Omusati region is close Etosha pan as seen in Fig. 1 which is the discharge area for the whole basin. On the other hand, pans in Ohangwena are characterized by $CaMgHCO_3$ and $NaHCO_3$ water types. $CaMgHCO_3$ are common in recharge zones and normally infer fast infiltration through preferential paths or can result from dissolution of calcite and dolomite when they react with CO_2 rich water (Freeze and Cherry, 1979). $NaHCO_3$ are commonly formed as result of silicate weathering and cation exchange while mixed water type in the ephemeral river could indicate mixed water sources.

Since all the collected samples are plotting close to the LMWL and within the range of modern local precipitation, this implies that these aquifers in both regions were recharged under climatic conditions similar to modern climate. The water in the deep wells in Ohangwena and in Omusati wells have a slightly depleted source water (−8.7, −58.2) and (−8.6, −57.5) respectively. This could indicate that recharge to the aquifers which these wells are tapping from is limited to larger amount events while for the shallow wells also small amount events are contributing as the mean parent water source is slightly enriched (−7.8, −51.8). In addition, the deep wells are plotting closer to the LMLW this indicates that the component from recharge without significant evaporation is larger than for the shallow wells. Samples from the shallow (open funnel-shaped) hand-dug wells are enriched in $\delta^{18}O$ and $\delta^{2}H$ irrespective of their latitude or longitude and they are strongly affected by evaporation due to exposure to the atmosphere. Bittner (2006) highlighted that high evaporation rates in the CEB causes the drying up of pans and ephemeral rivers resulting in the precipitation of salt and increased salinity of the shallow aquifers, in particular in waterlogged areas and areas comprising a low permeable lithology. It was also established that hand-dug wells in Omusati are tapping from both the perched aquifer and the regional aquifer and it is sometimes difficult to determine from which aquifer each well is tapping as the regional aquifer has a relatively shallow water table.

Temporal variations within each region are different for both regions. In Ohangwena temporal patterns were mainly observed in chloride and isotopes. The decrease in chloride content and isotopes during rainy season indicate recharge events and visibly they occur at different times depending on the rainfall. It is interesting to notice that the recharge events occur in the same period even though the aquifers have different depths to water table. If one assumes they are recharged by same rainfall events and same mechanisms, it would be expected that recharge will be delayed for the deep wells because of the percolation distance of the water to reach the aquifer in comparison to the shallow wells. Therefore, the results that the shallow wells and deep wells in Ohangwena are recharged through different mechanisms, whereby the deep wells are recharged through fast preferential paths and shallow wells by diffuse recharge. In Omusati the change is in Chloride and isotopes is observed but not following a particular pattern, however TDS increases during dry season. This indicates dissolution of evaporites and evaporation as the dominant process that influences the change. It should be noted that the in both regions, the sampled wells are used for water abstraction, but there are no estimation of how much is extracted and thus this could have an impact on accounting for temporal variations.

6 Conclusions

In the present study, it could be shown that it is indeed possible to clarify the hydrogeological complexities of heterogeneous perched aquifers using isotopic and hydrochemical tracers at different spatial and temporal scales. A distinct spatial difference between samples from Omusati and Ohangwena regions. The spatial heterogeneity as shown by the TDS, hydrochemical water types and chloride can be attributed to lithological and climatic factors. Deep wells are recharged by fast infiltration through preferential flow paths during large rainfall events while shallow wells can be recharged even by lighter events. The hydrochemical water types implies that pans in Ohangwena are recharge zones while in Omusati they are discharge zones. There are no piezometers and boreholes in these perched aquifers so far and thus the water level fluctuation data is missing. Therefore it's recommended, to establish monitoring stations for both water levels and water quality. This will also enable calculation of the current the abstraction rates from the perched aquifer.

Data availability. Individual data are available from the author(s) on request.

Author contributions. All authors were involved in the field work, interpretation of data and reviewing of the manuscript. PK was leading the hydrochemical analysis while HW analysed the stable isotopes. JTH prepared the manuscript with contributions from all co-authors.

Competing interests. The authors declare that they have no conflict of interest.

Special issue statement. This article is part of the special issue "Understanding spatio-temporal variability of water resources and the implications for IWRM in semi-arid eastern and southern Africa". It is a result of the IAHS Scientific Assembly 2017, Port Elizabeth, South Africa, 10–14 July 2017.

Acknowledgements. This work forms part of PhD study undertaken at the University of Namibia. This work was partly funded by the German Federal Ministry for Education and Research (BMBF) within the SASSCAL project (Southern African Science Service Centre for Climate Change and Adaptive Land Management) under contract number 01LG1201. The work also benefited from the National Commission of Research Science and technology (NCRST) grant under project code ncrst/phd/012. We thank the local communities' for access to sample their wells. We thank UNAM staff and local authorities who assisted in fieldwork. We kindly thank the staff at the BGR hydrochemical laboratory for analysing samples. Benjamin Mapani, your proof reading is highly appreciated.

Edited by: Hodson Makurira
Reviewed by: Richard Owen and one anonymous referee

References

Bittner, A.: Desk study report: Cuvelai-Etosha groundwater investigation. Windhoek, retrieved from: http://www.the-eis.com/data/literature/CuvelaiDeskStudyReport_Version1.1.pdf (last access: March 2018), 2006.

Freeze, R. A. and Cherry, J. A.: Groundwater, Prentice-Hall, Englewood Cliffs, USA, 1979.

Hamutoko, J. T., Wanke, H., and Voigt, H. J.: Estimation of groundwater vulnerability to pollution based on DRASTIC in the Niipele sub-basin of the Cuvelai Etosha Basin, Namibia, Phys. Chem. Earth, 93, 46–54, https://doi.org/10.1016/j.pce.2015.12.007, 2016.

Hamutoko, J. T., Wanke, H., Koeniger, P., Beyer, M., and Gaj, M.: Hydrogeochemical and isotope study of perched aquifers in the Cuvelai-Etosha Basin, Namibia, Isotopes Environ. Health Stud., 53, 382–399, https://doi.org/10.1080/10256016.2016.1273913, 2017.

Ledesma-Ruiz, R., Pasten-Zapata, E., Parra, R., Harter, T., and Mahlknecht, J.: Investigation of the geochemical evolution of groundwater under agricultural land: A case study in north-eastern Mexico, J. Hydrol., 521, 410–423, https://doi.org/10.1016/j.jhydrol.2014.12.026, 2015.

Lindenmaier, F., Miller, R., Fenner, J., Christelis, G., Dill, H. G., Himmelsbach, T., and van Wyk, B.: Structure and genesis of the Cubango Megafan in northern Namibia: implications for its hydrogeology, Hydro. J., 22, 1307–1328, https://doi.org/10.1007/s10040-014-1141-1, 2014.

Mendelsohn, J. M., Jarvis, A., and Robertson, T.: A profile and Atlas of the Cuvelai-Etosha basin: RAISON & Gondwana collection, Windhoek, Namibia, 2013.

Wanke, H., Nakwafila, A., Hamutoko, J. T., Lohe, C., Neumbo, F., Petrus, I., and Quinger, M.: Hand dug wells in Namibia: An underestimated water source or a threat to human health?, Phys. Chem. Earth, Parts A/B/C, 76–78, 104–113, https://doi.org/10.1016/j.pce.2015.01.004, 2014.

https://doi.org/10.5194/piahs-378-37-2018

Mapping groundwater availability and adequacy in the Lower Zambezi River basin

Blanca Pérez-Lapeña[1], **Francisco Saimone**[1], **and Dinis Juizo**[1,2]

[1]Salomon Lda., Maputo, Mozambique
[2]Civil Engineering, Eduardo Mondlane University, Maputo, Mozambique

Correspondence: Francisco Saimone (francisco.saimone@gmail.com)

Received: 10 November 2017 – Accepted: 11 December 2017 – Published: 29 May 2018

Abstract. Groundwater plays an important role as a source of water for various socio-economic uses and environmental requirements in the lower Zambezi basin in Mozambique. Hence it is important to know its availability and adequacy in space to inform decision making for sustainable water management practices. For a derivation of a Groundwater Availability map and a Groundwater Adequacy map we adapted the DRASTIC methodology in a GIS environment to determine how different parameters, such as precipitation, topography, soil drainage, land use and vegetation cover, aquifer characteristics and groundwater quality affect (i) groundwater recharge on a long-term sustainable basis, (ii) the short-term abstraction potential and (iii) the long-term adequacy of groundwater utilization for domestic use. Results showed that groundwater availability in the Zambezi basin varies mostly from medium to low, with highest potential along the perennial rivers and in the delta where it plays a crucial role in environmental preservation. The southern margin of the Zambezi River shows low groundwater availability and also presents low adequacy for domestic use due to poor groundwater quality. The results from this study will be used in determining the most promising future development pathways and select the most attractive strategic development plans of the Mozambican government for the Lower Zambezi basin.

1 Introduction

Groundwater is an essential source of water for various socio-economic uses in the lower Zambezi Basin (World Bank, 2009). It is also the only source of water supply to rural communities and the main source to two provincial capital cities in the center of the country, i.e. Quelimane (aquifer of Licuári) and Tete/Moatize (aquifers of the Nhartanda valley and those located on the banks of the Revúboè River). Groundwater also plays a crucial role in the preservation of valuable ecosystems in the basin. In semi-arid areas groundwater is an important resource for drought mitigation (Foster and Tyson, 2013). In contrast, recent trends of intensification of mining activities in the Zambezi valley bring several challenges for groundwater management, related to lowering of groundwater levels and the infiltration of acid and highly mineralized waters (Pondja et al., 2017).

It is within this context that it is necessary to evaluate the groundwater potential and the risks associated to its exploitation.

The presence, quantity, and quality of groundwater is related to geological, structural, climatic, morphological, surface drainage, chemical content, and land use/land cover factors and several models have emerged that seek to establish the groundwater potential in a basin (Ganapuram et al., 2009). This study followed a similar approach as the weighing method originally developed for groundwater contamination risk analysis (DRASTIC, Aller et al., 1987). Based on the work by SWECO (2004) and Plancenter (2008), an adaptation and extension of the method is presented to incorporate key parameters that affect groundwater recharge and its quality, and thus, indicate groundwater availability and adequacy. In the remaining of this paper we refer to groundwater as fresh water occurring in the saturated zone.

Published by Copernicus Publications on behalf of the International Association of Hydrological Sciences.

2 Methodology

2.1 Study area

The study area of this project covered the Zambezi river basin in Mozambique. The basin is the richest region in Mozambique in natural resources (due to its high potential in the agricultural, hydropower and mining sectors) as well as in tourism while the Delta is of high environmental importance. About four million people live within the boundaries of the Zambezi River basin, with the highest concentrations in the cities of Tete and Quelimane. Its climate varies from tropical wet in the Delta area, to tropical dry savanna in areas close to the Zambezi river and moderately humid, based on altitude, in the northern part of the basin. The general rainfall pattern shows a decrease from the coast to the most inland areas as a result of precipitation originating mostly from fronts and cyclones.

For the analyses, the basin was subdivided into 27 sub-basins that were defined considering the main confluences and sections of the most important tributaries and imposing maximum sub-basin areas on the order of 10 000 km^2 and minimum areas in the order of 2500 km^2 (Fig. 1).

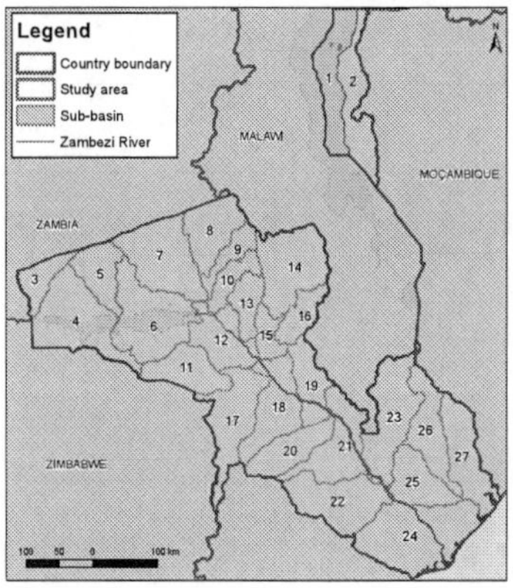

Figure 1. Study area and sub-basins.

2.2 The extended approach

The DRASTIC system is a widely used method to assess the vulnerability of groundwater resources to pollution. It is an overlay and index model designed to produce vulnerability scores by superimposition of environmental and geological thematic maps (Aller et al., 1987). In essence, the DRASTIC method is based on the hydraulic and hydrogeological characteristics of the aquifers. The DRASTIC method was adapted and extended in this study to evaluate the availability and viability of groundwater exploration. The following four categories and corresponding factors were considered in our approach:

1. the *depth to the water table* (D), affecting the potential exploitation of groundwater;
2. the *short-term abstraction potential* (P), governed by the aquifer material (A) and associated productivities (C);
3. the magnitude of groundwater *recharge* (R) on a long-term basis, governed by the mean annual precipitation (MAP), slope (T), soil infiltration (S), and land use and land cover (LU/LC);
4. *groundwater quality* (Q), affecting the adequacy of groundwater for domestic water supply.

Values in each factor were first divided into ranges. Each range was assigned a rate p on a scale 1 to 10 and then each category was assigned a weight w to quantify its relative importance in determining groundwater availability and adequacy for domestic use. Equation (1) was used to calculate the groundwater availability index and Eq. (2) was used to calculate the availability and adequacy index:

$$\text{Availability index} = D_p \cdot D_w + P_p \cdot P_w + R_p \cdot R_w \quad (1)$$

$$\text{Availability and adequacy index} = D_p \cdot D_w + P_p \cdot P_w + R_p \cdot R_w + Q_p \cdot Q_w. \quad (2)$$

The analysis was carried out in a GIS environment to produce spatially varying indices as presented in the groundwater Availability map and the map of Availability and Adequacy of groundwater use for the study area.

2.3 Hydrogeological categories

2.3.1 Depth to water table (D)

A database was constructed containing all identified spatial data on boreholes until November 2016 for the study area. Data on groundwater levels was obtained from 1028 boreholes (depth in m). The relationship between water depth values and the basin hydrogeology was first analysed by fitting a regression model to the (log-transformed) water depth data, using aquifer type as the dummy variable. After the analysis, we obtained a non-significant overall fit (at $\alpha = 0.05$). The depth to water table map was determined spatially by interpolating the water depth values using the Inverse Distance Weighing algorithm, IDW. In this study, rates were adopted from Aller et al. (1987) and are presented in Table 1 and applied to create the map in Fig. 2a.

2.3.2 Short-term abstraction potential (P)

The data contained in the Hydrogeological Map of Mozambique at scale of 1 : 1 000 000 was used to assess the short-term abstraction potential in the study area. The hydrogeological map contains information of (i) aquifer types and (ii) "yield classes" associated with each aquifer type. Types of aquifers, as described in the Hydrogeological map, are:

Table 1. Ratings for factors/categories.

Factor Depth to water (m)	Rate	Factor Slope (%)	Rate	Factor Land use and land cover	Rate
0–1.5	10	0–2	10	Irrigation cultivation	9
1.5–4.5	9	2–4	9	Raifed cultivation	8
4.5–9.1	7	4–7	8	Decidous shrubs; decidous	7
9.1–15.2	5	7–12	5	grassland (2–5 m high); rice	
15.2–22.8	3	12–18	3	cultivation, natural water bodies	
22.9–30.5	2	> 18	1	Everygreen shrubland; Everygreen	6
> 30.5	1			grassland (2–5 m high)	
Short-term abstraction potential		EC		Shrub savanna	5
Aquifer type A ($> 50\,m^3\,h^{-1}$)	10	0–700	10	Miombo open; Mopane open;	4
Aquifer type B ($> 50\,m^3\,h^{-1}$); type A ($10–50\,m^3\,h^{-1}$)	8	700–1500	8	Decidous forest; Open shrub on temporarily	
				flooded land; Shifting vegetation;	
Aquifer type A ($3–10\,m^3\,h^{-1}$), type B ($10–50\,m^3\,h^{-1}$)	6	1500–7500	3	Herbaceous vegetation temporarily flooded;	
				Evergreen forest; Tree crops; Tree savanna;	
Aquifer type C ($< 1\,m^3\,h^{-1}$)	3	> 7500	1	Woodland temporarily flooded land	
Soil infiltration		Recharge ($mm\,yr^{-1}$)		Galery forest; Miombo dense; Mopane dense;	3
Excessive	10	> 41.29	8	Semi-decidous open forest; Shifting cultivation;	
Little excessive; Good to excessive	9	26.48–41.29	6	Semi-evergreen (montaneous) forest;	
Good; Imperfect to good	8	20.67–26.78	4	Bare soils	
Imperfect	7	16.09–20.67	2	Built-up areas; Mangrove dense; Mangal open;	2
Moderate to imperfect	6	< 16.09	1	Semi-evergreen mountainous forest	
Imperfect to moderate	5			Bare Rocks	0
Moderate	4				
Moderate to bad; Imperfect to bad	2				
Bad to very bad; Lakes/lagoons/rivers	1				

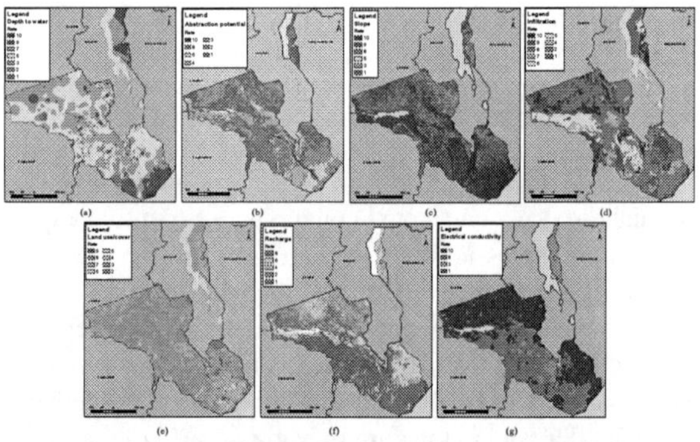

Figure 2. Rating of categories/factors: **(a)** depth to water table; **(b)** short-term abstraction potential; **(c)** slope; **(d)** soil infiltration; **(e)** land use/cover; **(f)** recharge; **(g)** groundwater quality (EC).

- Aquifer type A – intergranular aquifers (continuous and commonly unconsolidated);
- Aquifer type B – fissured aquifers (discontinuous and commonly consolidated);
- Aquifer type C – areas with local aquifers of limited productivity or areas without significant groundwater resources (intergranular or fissured).

Each aquifer type is associated with a range of "yield classes" from 1 to 3 where the lower number reflects higher short-term abstraction potential. The chosen rates are presented in Table 1 and applied to create the map in Fig. 2b.

2.3.3 Recharge (R)

Recharge is probably one of the most important parameters in groundwater planning and management. It is a hydrological process where water moves downward from surface water to groundwater and it gives an indication of the potential for exploitable productivities on the surface. The recharge estimate in this study was based on a weighted combination of the mean annual precipitation (MAP) map, the rated slope map, the rated soil map, and the rated land use/land cover map.

Slope

The slope map was constructed from the Digital Elevation Model (DEM), at a resolution of approximately 90 m, produced by the NASA Shuttle Radar Topography Mission (SRTM). To construct the rated slope map, flat terrain received a score of 10 (high positive effect on the amount of recharge), and lower scores were awarded whenever the slopes get steeper (lower effect on recharge). The chosen rates (adopted from SWECO, 2004 and Plancenter, 2008) are presented in Table 1 and applied to create the map in Fig. 2c.

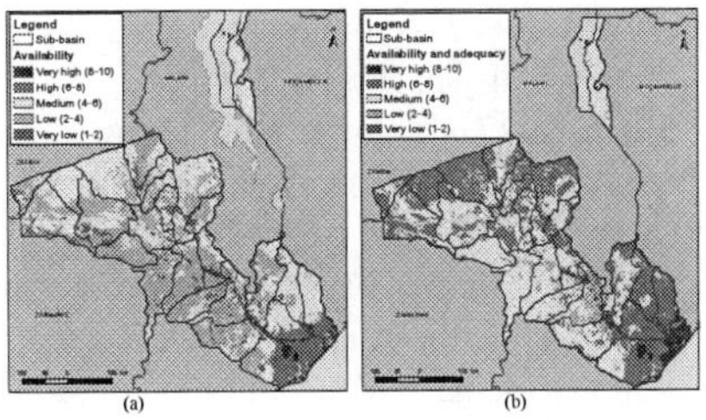
(a) (b)

Figure 3. **(a)** Availability Map and **(b)** Availability and Adequacy Map.

Soil infiltration

Information on soils was obtained from the Carta Nacional de Solos de Moçambique at a scale of 1 : 1 000 000. The dataset also contains information on the infiltration properties of the different soil types expressed as a categorical variable ranging from "very bad" to "excessive infiltration". To construct the rated soil map it was assumed that soils with a high infiltration have a high (positive) impact on the aquifer recharge. The chosen rates are presented in Table 1 and applied to create the map in Fig. 2d.

Land use/land cover

The land use and land cover map at a scale of 1 : 250 000 was obtained from the study on the Mozambican agroecological zoning. This factor has an important contribution to the potential of groundwater recharge. Artificial structures such as buildings and pavements of cement or asphalt reduce infiltration. Vegetation, on the other hand, can reduce the energy of the raindrops, which avoids soil compaction by impact force, thus favouring infiltration. Areas densely covered mainly by deeply rooted tree vegetation, promote evaporation at the expense of infiltration (Schilling et al., 2006). The rates, mostly based on literature (Preeja et al., 2011; Senanayake et al., 2016; Suganthi et al., 2013), are presented in Table 1 and applied to create the map in Fig. 2e.

Weighted estimation and rating

Groundwater recharge was calculated based on information on slopes, soils, land use/land cover and MAP in each sub-basin using a weighing procedure. In order to have greater control over the values obtained by the weighing procedure recharge thresholds were stablished for the study area. In this case, there were no previous detailed studies on the recharge potential of the aquifers in the Zambezi Basin. The closest reference is the Hydrogeological map of Mozambique which gives an indication of the value of recharge throughout the country. Thus, recharge thresholds were established based on the hydrogeological characteristics of the basin. In this study, we adopted the following thresholds, i.e. ceiling values for MAP (percMAP_n):

- type A aquifer: $\text{percMAP}_\text{A} = 16\,\%$;
- type B aquifer: $\text{percMAP}_\text{B} = 8\,\%$;
- type C aquifer: $\text{percMAP}_\text{C} = 4\,\%$.

The weighted estimate of recharge per sub-basin was calculated as follows (Eq. 3):

$$\begin{aligned}\text{Weighted estimate of recharge } &\left(\text{mm}\,\text{yr}^{-1}\right)\\ &= [(1\cdot\text{rated slopes}) + (2\cdot\text{rated soils})\\ &+ (3\cdot\text{rated LU/LC})]/60\cdot\text{percMAP}_n\cdot\text{MAP}. \quad (3)\end{aligned}$$

To validate the recharge estimates, the values obtained in this model were compared with the mean annual runoff (MAR) in each sub-basin, which was estimated using the Pitman model. It was therefore assumed that part of the estimated surface flow refers to the recharge of the aquifers in the basin. In general, MAR values tended to be low with a mean percentage of MAR to MAP over the sub-basins of 10.5 % (Wright, 1992). This value was in line with expected values reported in the scientific literature. At the same time, recharge values tended to be much lower, representing an average percentage of MAR to MAP of 3.4 %. This low value reflected well the fact that above 70 % of the basin is composed of base complex aquifers (type C) subject to low recharge.

The recharge estimates were rated according to a statistical analysis using the quantiles of their data distribution (Table 1). Figure 2f shows the recharge rated map.

2.3.4 Groundwater quality

To characterize groundwater quality in the Lower Zambezi basin, we used data on electrical conductivity (EC). We chose EC because it is the only parameter most frequently measured in the existing boreholes in the basin and it is therefore more expressive to characterize the groundwater quality at the basin level. The EC map was constructed by interpolating (using IDW) the EC values taking into account the information of the different aquifer types in the basin. EC data for the Niassa Province were not available; therefore this information was obtained from the hydrogeological map and integrated in the EC interpolation.

Ratings of EC were based on fitness for use thresholds (Table 1). These were taken to be $700\,\mu\text{S}\,\text{cm}^{-1}$ (upper limit for domestic consumption), $1500\,\mu\text{S}\,\text{cm}^{-1}$ (irrigation), and $7500\,\mu\text{S}\,\text{cm}^{-1}$ (livestock watering). The resulting rated EC map is shown in Fig. 2g.

3 Results

The final step in producing the map indicating groundwater availability was the use of a weighted sum of the "Depth to water table rated map" (after application of a weighting factor of 5), the "Recharge rated map" (weighting factor of 3.5), and the "Short-term abstraction potential rated map" (weighing factor 3). The groundwater availability map is shown in Fig. 3a. To generate the map indicating groundwater availability and adequacy, a weighted sum was used of the three abovementioned maps (with the corresponding weights) and the "Groundwater quality rated map" (weighing factor of 6). The groundwater availability and adequacy map is shown in Fig. 3b. The weights for the categories "Recharge", "Short-term abstraction potential", and "Groundwater quality" were slightly adjusted compared with previous studies in Mozambique (SWECO, 2004 and Plancenter, 2008) due to the inclusion of category "Depth to water table", which weight was adopted from the DRASTIC method (Aller et al., 1987).

To summarize the results by sub-basin obtained from the "Availability Map" and the "Availability and Adequacy Map", the average (area-weighted) score for each sub-basin was calculated and classified from very low (1–2) through very high (8–10). In general, water availability in the basin was medium to low, with the exception along the lower river margins and in the delta region. The southern margin of the river had on average the least potential, having sub-basins located in zones of low productivity and a corresponding low recharge. The quality of water in these basins was relatively low, possibly due to the occurrence of Sena arkosic sandstones. The northern left bank of the Zambezi River (sub-basins number 3, 5, and 7 in Fig. 1) and the sub-basin affluent to Lake Niassa (number 2) had medium availability of groundwater. However, the results for the latter sub-basin were less reliable due to poor data availability. Sub-basins located along the main river course presented on average low groundwater availability. More groundwater was available after the Cahora Bassa dam (sub-basin number 12), and further downstream (sub-basins number 15 and 19). Sub-basins located in the delta (24, 25, 26, 27) had the greatest groundwater availability. In relation to the adequacy of groundwater (quality) for human consumption, the situation generally improved from medium to high. The sub-basins along the southern bank of the main river courses were those having a relatively lower availability and adequacy than the other sub-basins.

4 Conclusions and discussion

It was concluded that groundwater availability is relatively low in the Zambezi basin. The main reason is that, as calculated from the data in the Hydrogeological map, 77 % of the Zambezi Basin is composed of aquifers of limited productivity (aquifer type C). When considering the adequacy of groundwater for human consumption, the situation improved with a large number of sub-basins having a medium to high groundwater availability and adequacy, which were mostly located in the northern bank of the Zambezi river, and the delta region. The potential for groundwater exploitation along the southern bank of the river seemed rather low because there was only low groundwater availability and there were also water quality problems.

The results in this study largely relied on aquifer productivities obtained from the hydrogeological map of Mozambique dated from 1987. An analysis was also carried out to obtain the productivities based on, more recent, borehole data. However, given the quality of these data it was decided that a more intensive data validation procedure was required before using the borehole data for the estimation of aquifer productivities in the basin. Given the paucity of data and the possible limitations in the specificity of the DRASTIC method to the reality of the study area, an improved approach could be to modify the classical model (e.g. V-DRASTIC) or combine it with other developed models as in the entropy-weighted fuzzy DRASTIC method (Souleymane and Zhonghua, 2017).

A common problem is the definition of the weights to be assigned to the parameters that are used in determining the relative groundwater potential. There are many techniques that are being tested and proposed ranging from the use of statistical methods based on frequency ratios, the weight of evidence that a given characteristic is more relevant than the other, or the use of the Analytic Hierarchy Process (AHP) (Zeinolabedinia and Esmaeily, 2015). The AHP is considered an innovation that aims at removing subjectivity in selecting and assigning weights to the parameters; however, applying it in this technical sphere of analysis has proven not yet to be sufficiently matured to be adopted in this study. This study used a weighting method for importance (weight of evidence), using similar weights as those developed in a risk analysis method for groundwater contamination (DRASTIC), which has been developed to a great extent in this field of analysis and, therefore, reduces the risk of subjectivity in selecting the weights to be assigned to the parameters under analysis. With regard to the weights used to obtain the "Availability Map", greater weight was given to the "Groundwater depth" category because it has great significance in the prospecting for groundwater for domestic purposes in Mozambique, where most of the population is rural and uses boreholes fitted with manual pumps. The use of an AFRIDEV pump, most used in these projects, is limited for greater depths. Therefore, to introduce this technological condition in the availability, a greater relative weight was given to the water depth category. A lower weight was given the "Recharge" category, which is greater than the weight of the "short-term abstraction potential" given the importance of first identifying the areas with high recharge potential before assessing the viability of water use through abstraction infrastructures.

With regard to the weights used to obtain the "Availability and Suitability Map", the greater weight was given to the "Groundwater quality" category, as this category gives an indication of the suitability of the use intended for the water (in this case for human consumption).

The recent trends of intensification of mining bring a new reality and challenge in the management of groundwater related to excessive lowering of water levels and the infiltration of acid and supermineralized drainage waters that can reach the water table. The impacts of these new developments need to be assessed and monitored to avoid potential degradation of groundwater quality in the basin. In this study, we used EC to describe groundwater quality as there exists borehole data with good coverage for the whole basin. Data on groundwater quality is still very incipient and does not allow for a spatial view for other relevant quality parameters such as zinc, copper, lead, cadmium and arsenic in those areas where mineralization and mining activities take place.

Finally, the estimated groundwater abstraction potential (short and long-term) as derived from the aquifers' recharge estimates showed that the potential for exploitation of groundwater resources is still very large compared to current uses of the aquifers. Nonetheless, such potential may only be materialized with a proper and intensive control of groundwater use to avoid degradation of the aquifers and ensure groundwater supply on a sustainable basis.

Data availability. Data used in this paper were obtained from various sources including global open data sets. Global data sets include vegetation cover, land use, and a digital elevation model. There are, however, data sets belonging to public entities in Mozambique that can be obtained on request and at a cost, namely the Hydrogeological map of Mozambique at scale 1 : 100 000, rainfall data from the National Meteorological Institute, and soil types from the National Cartographic Centre. Complementary climatic data such as rainfall and temperature, and borehole information are available on request and free of charge from the National Directorate of Water and Resource Management (DNGRH). The resulting maps of this study will be publicly available at the end of the overall study that is part of through the public web site of DNGRH http://www.dngrh.gov.mz.

Competing interests. The authors declare that they have no conflict of interest.

Special issue statement. This article is part of the special issue "Understanding spatio-temporal variability of water resources and the implications for IWRM in semi-arid eastern and southern Africa". It is a result of the IAHS Scientific Assembly 2017, Port Elizabeth, South Africa, 10–14 July 2017.

Acknowledgements. This article was developed within the framework of an ongoing project "Elaboração do plano estratégico para a utilização e desenvolvimento da bacia hidrográfica do rio Zambeze". The authors gratefully acknowledge the funding provided by the National Directorate of Water and Resource Management – Mozambique and the contributions of all members of the project team.

Edited by: Hodson Makurira
Reviewed by: two anonymous referees

References

Aller, L., Bennett, T., Lehr, J. H., and Petty, R. J.: DRASTIC: A Standardized System for Evaluating groundwater Pollution Potential Using Hydrogeologic Settings, EPA/600/2-87/035, US EPA, Washington, D.C., 1987.

Foster, S. and Tyson, G.: Groundwater development opportunities and management responsibilities – the mission for African Basin Organisations, GWP/ANBO Policy Brief, Global Water Partnership, http://www.gwp.org (last access: January 2018), 2013.

Ganapuram, S., Kumar, G. T. V., Krishna, I. V. M., Kahya, E., and Demirel, M. C.: Mapping of groundwater potential zones in the Musi basin using remote sensing data and GIS, Adv. Eng. Softw., 40, 506–518, 2009.

Plancenter: Joint Maputo River basin water resources study – Moçambique, Swaziland and South Africa, EuropeAid/120802/D/SV/ZA, Groundwater Resources Assessment Report No. 5.1/2007, 2008.

Pondja, E. A., Persson, K. M., and Matsinhe, N. P.: Assessment of coal mine water in Moatize by static and leaching tests, Sustain. Water Resour. Manage., 3, 403–412, 2017.

Preeja, K. R., Joseph, S., Thomas, J., and Vijith, H.: Identification of groundwater potential zones of a tropical river basin (Kerala, India) using remote sensing and GIS techniques, J. Indian Soc. Remote Sens., 39, 83–94, 2011.

Schilling, K., Li, Z., and Zhang, Y.: Groundwater-surface water interaction in the riparian zone of an incised channel, Walnut Creek, Iowa, J. Hydrol., 327, 140–150, 2006.

Senanayake, I. P., Dissanayake, D. M. D. O. K., Mayadunna, B. B., and Weerasekera, W. L.: An approach to delineate groundwater recharge potential sites in Ambalantota, Sri Lanka using GIS techniques, Geosci. Front., 7, 115–124, 2016.

Souleymane, K. and Zhonghua, T.: A novel method of sensitivity analysis testing by applying the DRASTIC and fuzzy optimization methods to assess groundwater vulnerability to pollution: the case of the Senegal River basin in Mali, Nat. Hazards Earth Syst. Sci., 17, 1375–1392, https://doi.org/10.5194/nhess-17-1375-2017, 2017.

Suganthi, S., Elango, L., and Subramanian, S. K.: Groundwater potential zonation by remote sensing and GIS techniques and its relation to the groundwater level in the coastal part of the Arani and Koratalai River Basin, Southern India, Sci. Res. J., 17, 87–95, 2013.

SWECO: Development of the Pungwe River Basin joint IWRM strategy, in: Monograph Report Annex IV: Groundwater Resources, SWECO Project no. 1150447, 2004.

World Bank: Bacia do Rio Zambeze – Análise de Oportunidades de Investimento Multissectorial, Banco Mundial – Gestão de Recursos Hídricos – Região de África, Washington, 2009.

Wright, E. P.: The Hydrogeology of Crystalline Basement Aquifers in Africa, Geol. Soc. Lond. Spec. Publ., 6, 1–27, 1992.

Zeinolabedinia, M. and Esmaeily, A.: Groundwater Potential Assessment using Geographic Information Systems and AHP Method (Case Study: Baft City, Kerman, Iran), ISPRS Archives, XL-1-W5, 769–774, 2015.

https://doi.org/10.5194/piahs-378-43-2018

Impacts of landcover changes on streamflows in the Middle Zambezi Catchment within Zimbabwe

Webster Gumindoga[1], **Hodson Makurira**[1], **and Bezel Garedondo**[1,2]

[1]Department of Civil Engineering, University of Zimbabwe P.O Box MP167 Mt. Pleasant, Harare, Zimbabwe
[2]Department of Irrigation Development, Gwanda, Zimbabwe

Correspondence: Webster Gumindoga (wgumindoga@gmail.com)

Received: 17 November 2017 – Revised: 9 April 2018 – Accepted: 1 May 2018 – Published: 29 May 2018

Abstract. We investigate the impacts of land cover changes on the river flows of the Middle Zambezi tributary catchments in Zimbabwe. Trend analysis on rainfall and streamflow was carried out using the Mann-Kendall test at monthly and annual time steps. Rainfall analysis indicated an increasing trend which was not statistically significant ($p < 0.05$) for all stations. Annual streamflow time series indicated negative decreasing trends which were not statistically significant ($p < 0.05$) except for the rainfall months of November and December. The study deduced that the changes in rainfall did not affect hydrological catchment behaviour and changes in streamflow were thus caused by anthropogenic factors such as land cover changes. Statistical tests indicated a weak but significant correlation between rainfall and streamflow which also supports the fact that changes in streamflow are mainly driven by land cover changes. Land cover change assessments were done through supervised classification of Landsat images for the years 1989, 1998, 2008 and 2014. All catchments exhibited increases in cultivation area and decreases in forest and grassland. The semi-distributed HBV-Light model was applied for change detection modelling of the gauged Musengezi catchment. We conclude that the HBV Light model can be successfully used to simulate flows for the catchment.

1 Introduction

The Zambezi River Basin is the fourth largest African freshwater catchment and the largest river system in Southern Africa covering an area of 1.37 million km^2 with a discharge averaging around $2600\,m^3\,s^{-1}$ into the Indian Ocean (Beilfuss, 2012). The population of the basin is estimated at over 30 million (World Bank, 2010). Approximately 7.5 million (25 %) live in the main urban centres and the remainder in rural areas (ZAMCOM/SADC/SARDC, 2015). The expected increase in population, hence increase in consumptive use, will be coupled with land use changes that may, in turn, have an impact on the flow regimes of the basin. The Middle Zambezi part of the basin in Zimbabwe is also experiencing these increased pressures including increasing demands for water. Besides population increase, changes in agricultural practices and deforestation have been witnessed. It is necessary to investigate to what level these environmental pressures influence river flow changes and, also, to propose measures to mitigate negative impacts on river flow regimes.

Land use and land cover change drives many environmental models such as hydrological models within large and smaller river basins. The recent interests towards integrated management of land and water has renewed the need of quantifying the hydrological impacts of land use changes (Gumindoga et al., 2014) The integrated management of land and water resources will facilitate more productive and sustainable utilisation of the natural resources available in the basin.

This study seeks to understand and quantify the impacts of land cover changes on river flows in the Middle Zambezi Basin so as to devise ways of mitigating their impacts. GIS and remote sensing techniques will be integrated into a semi-distributed hydrological model to evaluate the historical impacts. This improved understanding on hydrology, land use management, land cover change interactions is critical for improved water and ecosystem management.

Published by Copernicus Publications on behalf of the International Association of Hydrological Sciences.

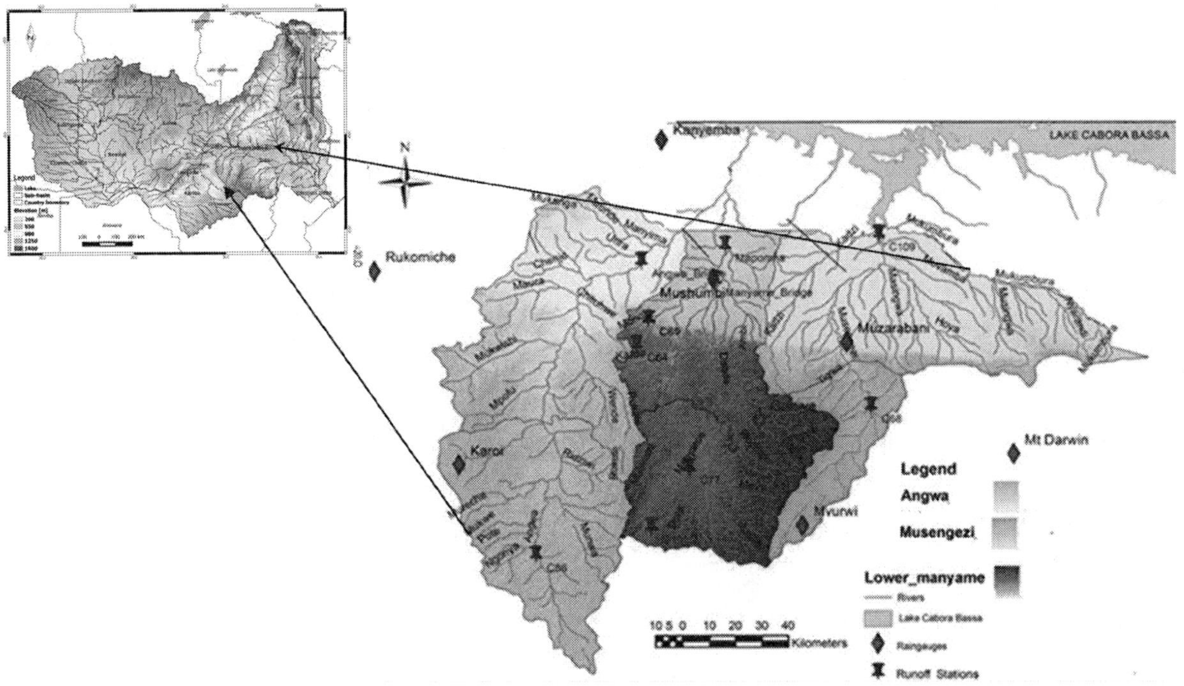

Figure 1. Zambezi River Basin with sub basins, major lakes, rivers, elevation. The location of Angwa, Lower Manyame and Musengezi Catchments from the Zambezi Basin is also shown.

Table 1. Runoff stations in study area.

Catchment	Station Name	Code	Subzone	Period Avaible		No. of years availabe
				From	to	
Angwa	Angwa-Chengu	C86	CA2	1975	1994	19
Lower Manyame	Manyame-nyakupunu	C64	CH2	1993	2013	20
	Manyame-Yomba Manayame-Balancing	C77	CH2	1988	2011	23
	Dam	C101	CH2	1989	2010	21
Musengezi	Musengezi	C68	CUG2	1966	2011	43

2 Study area

The Middle Zambezi Basin stretches from Victoria Falls to Lake Cabora Bassa. On the Zimbabwe side it has many tributaries and amongst them are the Angwa, Lower Manyame and Musengezi catchments all these draining directly into Lake Cabora Bassa (Fig. 1). Only Manyame River is perennial, with the rest being ephemeral. These three river systems are part of the Manyame Catchment of Zimbabwe. The Angwa catchment has a total area of 9479 km^2, and is made up of two subzones CA1 and CA2 according to the ZINWA subzone demarcations. The Lower Manyame catchment, comprising of two sub zones CH1 and CH2, has a total area of 6308 km^2. The Musengezi catchment with subzones CUG1 and CUG2 has a total area of 6925 km^2. All, except for the Manyame, are independent of upstream flow contributions.

The lower parts of these catchments lie in the Zambezi Valley in the Mbire and Muzarabani districts and experience flash floods which are partly a result of backwater effects when the downstream Lake Cahora Bassa records high storages as well as flows from upper catchments of the main rivers (Matingo, 2017). Average daily temperatures in the three catchments range from 24 to 32 °C, with the highest temperatures recorded on the northern parts of the catchments in the Zambezi valley reaching as high as 40 °C. The Mbire and Muzarabani districts receive low rainfall of approximately 450–650 mm year^{-1} and is classified as one of the driest agro ecological regions of Zimbabwe.

3 Material and methods

3.1 Data acquisition

3.1.1 Hydro-meteorological data

Runoff data was obtained for four stations (C86, C64, C77, C77, C101 and C68) in the lower Manyame, Musengezi and Angwa catchments. The stations have varied lengrhs of records ranging between 19 and 43 years. The details of runoff and rainfall stations are outlined in Table 1.

3.1.2 Satellite images

Freely available Landsat images were acquired from https://earthexplorer.usgs.gov/ (last access: 23 May 2018). The data was used for land use and land cover change mapping for Angwa, Lower-Manyame and Musengezi catchments. Landsat TM, ETM+ and Landsat 8 images were acquired between the years 1989 and 2014 which had approximately similar rainfall and distribution for each study catchment. The images were for the same season between June and September being the dry season for the three catchments. During the dry season, chances of obtaining minimum cloud cover of, preferably 0 %, are high. In addition, during this cool dry season, grass would have senesced, while the crops would have been removed from the fields thus allowing for better classification of land cover. The Angwa catchment is covered by four tiles with path/row combination 170/071, 170/072, 171/071 and 171/072. The lower Manyame catchment is covered by 2 tiles path/row combination 170/071 and 170/072. The path/row combination for the 3 tiles covering Musengezi catchment is 170/071, 170/072 and 169/071.

3.1.3 Digital Elevation Data

In hydrologic modelling, high-resolution spatially distributed elevation data is required. The ASTER (30 m resolution) DEM required for DEM hydro-processing (catchment extraction) in ILWIS Software was downloaded from https://earthexplorer.usgs.gov/ (last access: 23 May 2018).

3.1.4 Evapotranspiration

Evapotranspiration was derived from Moderate Resolution Imaging Spectroradiometer (MODIS 16) satellite. The data was obtained for monthly time steps and was then processed in ILWIS to get the monthly average evapotranspiration over the gauged part of Musengezi catchment.

3.2 Trend analysis in rainfall and stream flow

The hydrological functions of a catchment can be altered by climatic factors and human interferences which can be evaluated through trend analysis of hydro-meteorological records (Rientjes et al., 2011). The study determined trends in streamflows at stations C68, C86, C77 and C101. Rainfall data was obtained for Guruve, Muzarabani, Kanyemba, Karoi, Rukomiche, Mvurwi and Mt. Darwin stations.

The data was tested for normally distribution, stationarity, consistency and homogeneity as described in Dahmen and Hall (1990). The hydro-meteorological data was visually screened perfore perfoming normality tests, plausibility checks, data consistency and homogeneity tests and trend analyses (Uhlenbrook et al., 2010). The Mann–Kendall (MK) test, a non-parametric rank based method was used to test for trends in the streamflow records and long-term rainfall (Mann, 1945; Kendall, 1975) as given in Eq. (1):

$$= \frac{\sum_{i=1}^{n-1}\sum_{j=i+1}^{n} sgn(Y_j - Y_i)}{\delta_s} \quad (1)$$

With standard deviation:

$$\delta_s = \sqrt{\frac{n(n-1)(2n+5) - \sum_{i=1}^{n} t_i (i-1)(2i+5)}{18}} \quad (2)$$

Where:

- $sgn(Y_j - Y_i) = \begin{cases} 1, & \text{if} \quad Y_j - Y_i > 0 \\ 0, & \text{if} \quad Y_j - Y_i = 0 \\ -1, & \text{if} \quad Y_j - Y_i < 0 \end{cases}$
- n is the number of data
- Y_j and Y_i are the data values in two consecutive periods
- t_i is the number of ties, i.e. equal values, of extent i and n is the number of tied groups.

The test statistic S is asymptotically normal with zero mean and standard deviation σ_s; as computed from Eq. (3) (Hirsch and Slack, 1984). For cases where the sample size n is larger than 10, the standard normal variate p is computed by using the following equation (Rientjes et al., 2011).

$$p = \begin{cases} \frac{S-1}{\delta_s} & \text{if} \quad S > 0 \\ 0 & \text{if} \quad S = 0 \\ \frac{S-1}{\delta_s} & \text{if} \quad S < 0 \end{cases} \quad (3)$$

Upward trends are characterised by a positive value of S, whilst downward trends signal a negative value (Partal and Kahya, 2006).

3.3 Change detection in rainfall and streamflow

The year at which there was a change in stream flow was identified using the moving average t test (Rientjes et al.,

2011; Zheng et al., 2007). The test statistic of the t test is given by:

$$t = \frac{\bar{x}_1 - \bar{x}_2}{s\sqrt{\frac{1}{N_1} - \frac{1}{N_2}}} \tag{4}$$

and

$$s = \sqrt{\frac{N_1 s_1^2 + N_s s_2^2}{N_1 + N_2 - 2}} \tag{5}$$

Where:

- x_1 and x_2 = mean annual stream flow of the N years before the potential change point and the N years after the potential change point, respectively;
- N_1 and N_2 = the number of years in the first and second period;
- s_1 and s_2 = standard deviations in the first and second periods;
- s = pooled standard deviation.

N_1 and N_2 was equal for this study. Flow indices in each period were estimated, after having identified periods where mean annual flows are significantly different. These statistics include a high flow index (Q_5/Q_{50}) and a low flow index (Q_{95}/Q_{50}), where Q_k is the flow with probability of exceedance of k % of the time. The Weibull plotting position formula was applied to estimate exceedance probabilities as the approach relates to minima (i.e. smallest extreme values) as compared to Gumbel and Fréchet formula, for instance, that relate to maxima (i.e. largest extreme values).

3.4 Image classification for Land use/cover Changes

Processing of the downloaded images involved image rectification or geo-referencing and image enhancement. Image enhancement was done by preparing a colour composite of Landsat bands 5, 4 and 3 for Landsat TM images and bands 6, 5, 4 for Landsat 8 images (images visualised in their true colour) for each study year and catchment. The ILWIS software was used to perform image analysis and classification to obtain the land cover map. Ground truthing was achieved using a handheld GPS where a total of 150 points corresponding to the different landcover classes were taken for accuracy assessment.

3.5 Determining the impacts of land use and land cover changes on the streamflows using the HBV Light Model

The HBV light model (Seibert and McDonnell, 2010) was applied to examine the time series of model residuals which basically means calibration for a period before (or after) change and simulation of runoff that would have been observed without land-cover changes (reconstruction of natural runoff series) following Seibert and McDonnell (2010). Runoff was simulated at a daily timesteps with daily precipitation and temperature and monthly long term potential evapotranspiration as input variables to the model (Seibert and Vis, 2012). Areal rainfall and temperature time series were interpolated for different elevation zones. The semi-distributed model structure was adopted with three vegetation zones for each of the 6 elevation zones. The classified land cover maps for the year 2008 were then used to define vegetation zones in each catchment which include cultivation, water and forest. The DEM was used for estimation of catchment boundaries and elevation.

Daily discharge data obtained from ZINWA for gauging station C68 was also used as input into the model as well as average daily rainfall and mean temperatures of the surrounding stations (Mvurwi, Guruve and Muzarabani) obtained from the Meteorological Office.

3.6 Model Calibration and validation

Model calibration (October 1998 to September 2006) was achieved by use of Monte Carlo simulations (MCs) procedure, where best performing parameter sets were randomly generated from various model simulations. Model parameters were then manually adjusted, in order to obtain a process-based representation of the hydrological characteristics in the study catchments (Uhlenbrook et al., 2010). The objective of MCS is to explore the entire parameter space to find the best performing parameter sets. The Nash–Sutcliffe efficiency (Nash and Sutcliffe, 1970) (NSE) and the relative volumetric error (RVE) were used to assess the model efficiency. Model validation was done for the period October 2007 to September 2008.

3.7 Model application for land cover change impacts

The land cover map for the year 2008 was used in calibrating the model and obtaining of optimal parameter sets as available meteorological data was up to the year 2008. Parameters were calibrated for a reference period and the simulated runoff compared to observed runoff in periods with potential change (Seibert and McDonnell, 2010). To detect if there has been any changes as a result of land use, all the optimized parameters and the input variables derived from running the model for the period 2000 to 2008 were held constant as in Gumindoga et al. (2014), then the land cover parameters were varied to 1989, 1998 and 2014 ones and the model simulated again to observe if there were any differences to the runoff simulated In the model setup, only the catchment characteristics derived from land cover maps were changed, the precipitation, temperature and evapotranspiration values which were applied to simulations for the period 1998 to 2008 remained unchanged thus model forcing re-

Table 2. The Mann Kendal test statistic (Test Z) for trend analysis for monthly and annual rainfall for Mvurwi, Guruve and Muzarabani and the Sen's slope estimate (Q).

Time Series	Mvurwi (1970–2008)				Guruve (1970–2013)				Muzarabani (1991–2013)			
	n	Test z	Sig	Q	n	Test z	Sig	Q	n	Test z	Sig	Q
ANNUAL	38	−1.207		−5.104	40	+0.198		0.451	23	+0.106		1.065

The "+" sign means an increasing trend and the "−" sign means a decreasing trend. *: statistically significant at 5 % level.

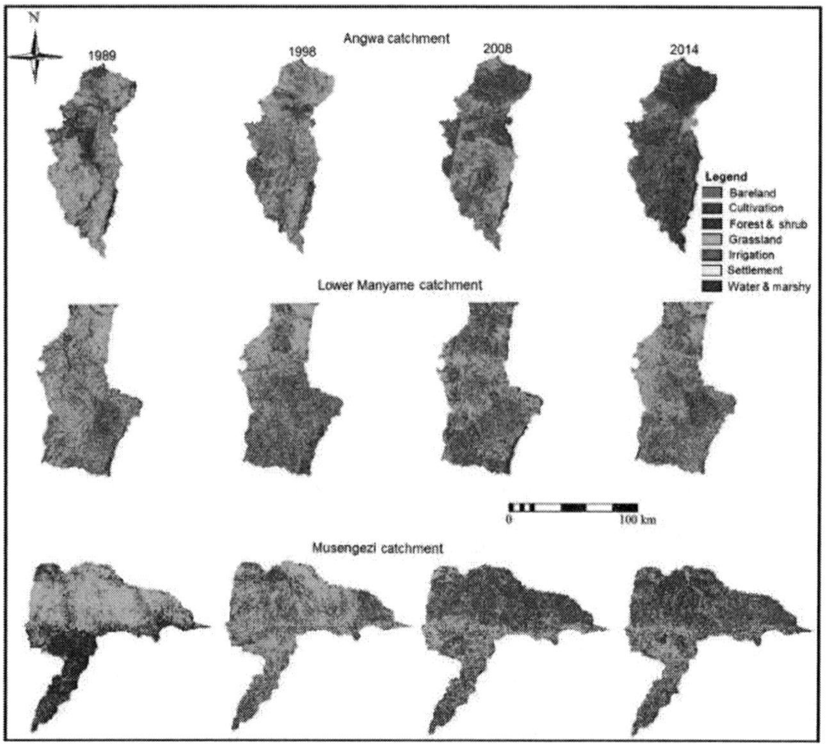

Figure 2. Middle Zambezi catchment land cover classification results for the years 1989, 1998, 2008 and 2014.

mained unchanged making the effect of land cover change explicit in the model though this maybe an over simplification.

4 Results and discussion

4.1 Rainfall trend analysis

Results of the trend test for Mvurwi, Guruve and Muzarabani indicated that there was a decreasing trend in the monthly received rainfall for September, October and December to July (Table 2). The month of May is the only month that had a statistically significant decreasing trend at 5 % level of significance.

The Sen's estimator (Q) for true slope of linear trend showed negative slopes for seven months of the year that include the rainfall season months for Zimbabwe and a positive slope for November. The dry season periods of June to September showed zero for the Sen's slope.

Similarly, trends for Kanyemba station were not significant for all months while Rukomiche station showed an increasing trend for months of December, January and March which are all in the wet season.

Trends for Karoi station were found to be statistically insignificant for all months of the year. All the trends for Mt Darwin station were not significant.

4.2 Streamflow trend analysis

Analysis of annual streamflow for station C68 located in Musengezi catchment showed a negative trend that is not statistically significant ($p = 0.05$). Analysis on monthly streamflows showed negative, downward trends for eleven months of the year and a positive trend for February only. Months in

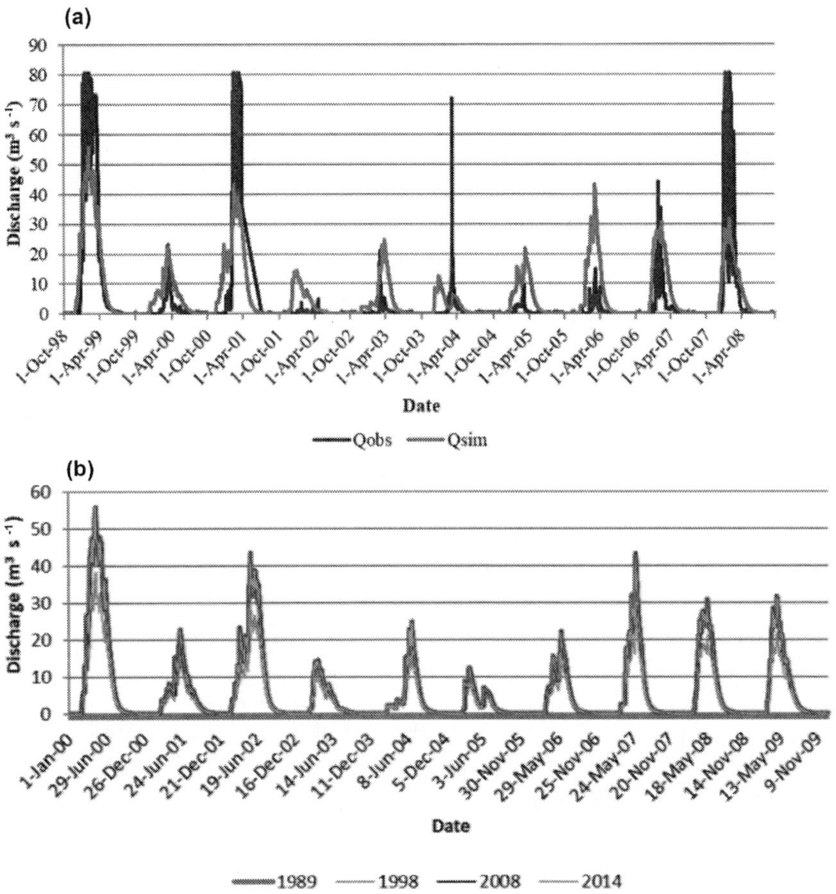

Figure 3. (**a**) Observed and simulated streamflow at C68 (1998–2008) and (**b**) streamflow hydrograph comparisons under the four land cover change periods.

the dry season between May and August showed statistically significant negative downward trends.

Statistically significant trends were also indicated for the months of November and December which all fall within the rainy season. Streamflows at station showed three clear segments where change of trend was detected. The mean annual discharge was $3.26\,m^3\,s^{-1}$ for the period 1967 to 1972, then increased to $4.97\,m^3\,s^{-1}$ for the period 1976 to 1983 and decreased to $4.74\,m^3\,s^{-1}$ between 1995 to 2011.

Monthly streamflows at C101 in Lower Manyame Catchment showed significant negative, downward trends for the months of July, August and November while trends for all the other months were not significant at $p = 0.05$ with June and September showing negative trends and the rest of the months showing positive increasing trends.

4.3 Results of image classification

Land cover changes were analysed for the period 1989, 1998, 2008 and 2014 for the three study catchments of Angwa, Lower Manyame and Musengezi. All catchments exhibited increases in cultivation area and decreases in forest and grassland (Fig. 2). The overall accuracy for Angwa catchment, Lower-Manyame and Musengezi Catchment is 88.57, 85.71 and 86.54 % respectively. Results of the overall accuracies are all above 85 % which means the images represent well the land cover features on the ground and are therefore suitable for further use. Highest accuracies were obtained for Bareland, Water and Marshy, Cultivation and Irrigation.

4.4 Model calibration and validation results

For calibration period, a reasonably good model performance with a Nash Sutcliffe Efficiency (NSE) of 0.61 was obtained using the land cover map of 2008 which were considered as the baseline conditions. A Relative Volume Error (RVE) and Percentage Bias (PBIAS) of 7.2 and 8.5 % respectively were obtained which as well indicates a reasonable performance. The model underestimated the peak flows and overestimated the low flows. The model validation was for the period 2006 to 2008. The model reproduced the observed streamflow hydrograph but under estimated the peak flows as a result underestimating the total observed streamflows. A NSE of 0.57 was obtained with a RVE and PBIAS of 8.9 and 8.7 % respectively indicating a reasonable model performance as well.

4.5 Simulation results under land cover changes

Simulation results for the period 1998 to 2008 served as the reference to simulations of the historic and future periods (Fig. 2: top). The simulations followed the pattern in the observed streamflow. However, it was observed that peak flows were under estimated and low flows over estimated. The simulated hydrographs represented the baseflows well. Figure 3b shows the streamflow hydrograph comparisons under the four land cover change periods.

5 Conclusions and recommendations

1. Increasing but insignificant trends were observed in rainfall received over the studied catchments while decreasing insignificant trend were observed in streamflows. This strongly suggests that the changes observed in streamflows are not directly caused by changes in rainfall but other drivers such as, possibly, land use changes.
2. Angwa, Lower Manyame and Musengezi have undergone notable land use and land cover changes over time.
3. Change detection modelling results indicated that changes in streamflows are caused by land use and land cover changes as was indicated by the changed simulation results of the mean flows and peak flows between the period 1989 and 1998 to 2014.

We recommend for policies that curb massive human induced deforestation that has been affecting the Middle Zambezi Basin. The cultivation of flood-prone land should also be prohibited. The regular analysis of land cover change on hydro-meteorological time series is urgently needed to ascertain the scales of changes from time to time. Integrated land use plans should be developed and implemented. Future studies should estimate actual evapotranspiration to evaluate how the basin water balance is affected by deforestation and land use changes.

Data availability. Data are not publicly accessible due to restrictions imposed by the supplier of the data.

Competing interests. The authors declare that they have no conflict of interest.

Special issue statement. This article is part of the special issue "Understanding spatio-temporal variability of water resources and the implications for IWRM in semi-arid eastern and southern Africa". It is a result of the IAHS Scientific Assembly 2017, Port Elizabeth, South Africa, 10–14 July 2017.

Acknowledgements. The authors greatly appreciate the DANIDA Fellowship under the WaterNet Phase 3 for funding this research.

Edited by: Jean-Marie Kileshye-Onema
Reviewed by: two anonymous referees

References

Beilfuss, R.: A Risky Climate for Southern African Hydro: Assessing hydrological risks and consequences for Zambezi River Basin dams, International Rivers Network (IRN), 60 pp., https://doi.org/10.13140/RG.2.2.30193.48486, 2012.

Dahmen, E. and Hall, M. J.: Screening of hydrological data: tests for stationarity and relative consistency, 49, International Institute for Land Reclamation and Improvement, Wageningen, the Netherlands, 1990.

Gumindoga, W., Rientjes, T., Shekede, M. D., Rwasoka, D. T., Nhapi, I., and Haile, A. T.: Hydrological Impacts of Urbanization of Two Catchments in Harare, Zimbabwe, Remote Sens.-Basel, 6, 12544–12574, 2014.

Hirsch, R. M. and Slack, J. R.: A nonparametric trend test for seasonal data with serial dependence, Water Resour. Res., 20, 727–732, 1984.

Kendall, M.: Rank correlation measures, Charles Griffin, London, 202, 1975.

Mann, H. B.: Nonparametric tests against trend, Econometrica, 13, 245–259, 1945.

Matingo, T.: Evaluation of sub daily satellite rainfall estimates through flash flood modelling in the Lower Middle Zambezi Basin, MSc Thesis, University of Zimbabwe, Department of Civil Engineering, 2017.

Nash, J. and Sutcliffe, J.: River flow forecasting through conceptual models part I – A discussion of principles, J. Hydrol., 10, 282–290, 1970.

Partal, T. and Kahya, E.: Trend analysis in Turkish precipitation data, Hydrol. Process., 20, 2011–2026, 2006.

Rientjes, T. H. M., Haile, A. T., Kebede, E., Mannaerts, C. M. M., Habib, E., and Steenhuis, T. S.: Changes in land cover, rainfall and stream flow in Upper Gilgel Abbay catchment, Blue Nile basin – Ethiopia, Hydrol. Earth Syst. Sci., 15, 1979–1989, https://doi.org/10.5194/hess-15-1979-2011, 2011.

Seibert, J. and McDonnell, J. J.: Land-cover impacts on streamflow: a change-detection modelling approach that incorporates parameter uncertainty, Hydrolog. Sci. J., 55, 316–332, 2010.

Seibert, J. and Vis, M. J. P.: Teaching hydrological modeling with a user-friendly catchment-runoff-model software package, Hydrol. Earth Syst. Sci., 16, 3315–3325, https://doi.org/10.5194/hess-16-3315-2012, 2012.

Uhlenbrook, S., Mohamed, Y., and Gragne, A. S.: Analyzing catchment behavior through catchment modeling in the Gilgel Abay, Upper Blue Nile River Basin, Ethiopia, Hydrol. Earth Syst.

Sci., 14, 2153–2165, https://doi.org/10.5194/hess-14-2153-2010, 2010.

World Bank: The Zambezi River Basin: A Multi-Sector Investment Opportunities Analysis – Summary Report, World Bank, available at: https://openknowledge.worldbank.org/handle/10986/2958 (last access: 23 May 2018), 2010.

ZAMCOM/SADC/SARDC: Zambezi Environment Outlook 2015, Harare, Gaborone, 2015.

Zheng, H., Zhang, L., Liu, C., Shao, Q., and Fukushima, Y.: Changes in stream flow regime in headwater catchments of the Yellow River basin since the 1950s, Hydrol. Process., 21, 886–893, https://doi.org/10.1002/hyp.6280, 2007.

https://doi.org/10.5194/piahs-378-51-2018

A GIS-based estimation of soil erosion parameters for soil loss potential and erosion hazard in the city of Kinshasa, the Democratic Republic of Congo

Martin Tshikeba Kabantu[1], Raphael Muamba Tshimanga[1], Jean Marie Onema Kileshye[3,2], Webster Gumindoga[3], and Jules Tshimpampa Beya[1]

[1]Dept. of Natural Resources Management and the Congo River basin network for research and capacity building in water resource (CB-HYDRONET), University of Kinshasa, P.O. Box 117, Kinshasa XI, Kinshasa, Democratic Republic of the Congo
[2]WaterNet, P.O. Box MP 600 Mount Pleasant, Harare, Zimbabwe
[3]Dept. of Civil Engineering, University of Zimbabwe, P.O. Box MP167 Mount Pleasant, Harare, Zimbabwe

Correspondence: Martin Tshikeba Kabantu (madeleinekabantu@gmail.com)

Received: 12 November 2017 – Revised: 11 April 2018 – Accepted: 13 April 2018 – Published: 29 May 2018

Abstract. Soil erosion has detrimental impacts on socio economic life, thus increasing poverty. This situation is aggravated by poor planning and lack of infrastructure especially in developing countries. In these countries, efforts to planning are challenged by lack of data. Alternative approaches that use remote sensing and geographical information systems are therefore needed to provide decision makers with the so much needed information for planning purposes. This helps to curb the detrimental impacts of soil erosion, mostly emanating from varied land use conditions. This study was carried out in the city of Kinshasa, the Democratic Republic of Congo with the aim of using alternative sources of data, based on earth observation resources, to determine the spatial distribution of soil loss and erosion hazard in the city of Kinshasa. A combined approach based on remote sensing skills and rational equation of soil erosion estimation was used. Soil erosion factors, including rainfall-runoff erosivity *R*), soil erodibility (*K*), slope steepness and length (*SL*), crop/vegetation and management (*C*) were calculated for the city of Kinshasa. Results show that soil loss in Kinshasa ranges from 0 to 20 $\mathrm{t\,ha^{-1}\,yr^{-1}}$. Most of the south part of the urban area were prone to erosion. From the total area of Kinshasa (996 500 ha), 25 013 ha (2.3 %) is of very high (> 15 $\mathrm{t\,ha^{-1}\,yr^{-1}}$) risk of soil erosion. Urban areas consist of 4.3 % of the area with very high (> 15 $\mathrm{t\,ha^{-1}\,yr^{-1}}$) risk of soil erosion compared to a very high risk of 2.3 % (> 15 $\mathrm{t\,ha^{-1}\,yr^{-1}}$) in the rural area. The study shows that the soil loss in the study area is mostly driven by slope, elevation, and informal settlements.

1 Introduction

The Soil Conservation Society of America provides a definition of soil erosion as the wearing away of the land surface due to the action of several agents, including running water, wind velocity, and gravitation creep (Ghanshyam, 2012). It represents the gross amount of soil removed or dislocated by different agents of nature, thus resulting in soil loss. The later represents the net amount of soil removed off a particular area. Soil erosion is known to cause severe land degradation, loss of soil productivity, and sedimentation of waterways (Pimentel and Burgess, 2013; Bhattacharyya et al., 2015). The Erosion Hazard assessment determines whether the risk of soil erosion is low, medium or high. While, soil erosion is influenced by geomorphological processes (Telles et al., 2013), it is accelerated by anthropogenic factors that develop spontaneously, with little or no planning (Robertson, 2009). All this has detrimental impacts on socio economic situations and the environment, especially in developing countries, thus increasing poverty.

Since 1980s, the African continent has been subjected to urban expansion, due to population growth and search for better life, but all this does not commensurate with infrastructure development (Antoine, 1991). In the Democratic Republic of Congo (DRC), the political instability and civil conflicts that have been observed over the past three decades have contributed to massive movement of people from rural areas and those hotspot areas of instability to more stable areas such as the city of Kinshasa. This has led to rapid demographic increase, deterioration of existing infrastructure, and informal settlements. For instance, the city of Kinshasa that was initially built for less than one million people, has now more than ten million inhabitants (Kinshasa-CPSRP, 2011). The expansion of the urban areas of the city is neither supported with appropriate planning nor adequate infrastructure. This situation contributes to widespread and high level land degradation and soil erosion, which is even difficult to monitor and control (Daniel et al., 2015; Bizuwerk et al., 2004). Assessing soil loss potential and erosion hazard is therefore important for best practices of soil conservation and management to reduce devastating consequences to the bio-physical and socio-economic environment. Most of soil erosion estimation approaches use parameters that require input data at appropriate spatial and temporal scales. Such data include for instance factors of rainfall erosivity, soil erodibility, land cover and slope characteristics that are difficult to acquire in data scarce areas such as the city of Kinshasa. This is where earth observation techniques such as the use of Geographical Information Systems (GIS) and Remote Sensing can be of unprecedented benefit to determine the soil erosion parameters necessary to assess soil loss potential and erosion hazard, establish a model of spreading and provide adequate management practices. A study carried out by Kitambo (2017) in one of the peri-urban catchments of Kinshasa, the N'sele catchment, estimated about $92 \pm 11\,\mathrm{t\,km^{-2}\,yr^{-1}}$ of soil loss. Mfumu et al. (2012) estimated the soil loss at the nuclear facility of Kinshasa. The study revealed that the soil loss at that the nuclear facility varies from 0 to $9\,\mathrm{t\,ha^{-1}\,yr^{-1}}$. The above mentioned studies used the Erosion Potential Model approach (Mfumu et al., 2012).

The objective of this study is to determine the parameters of soil erosion necessary for assessing spatial distribution of soil loss and erosion hazard in the city of Kinshasa using alternative sources of data, based on earth observation resources.

2 Study Area

Figure 1 shows the location of the study area, the city of Kinshasa, that is located west of DRC between 3.9 and 5.1° Latitude South and between 15.2 and 16.6° Longitude East, and extend over $9965\,\mathrm{km}^2$. The city is composed of 24 municipalities with 22 in urban part and 2 in rural (Nsele and Maluku), the rural area covers more than 75 % of the city area (Kinshasa-CPSRP, 2011). The study area is under the code AW4 climate according to the Köppen classification

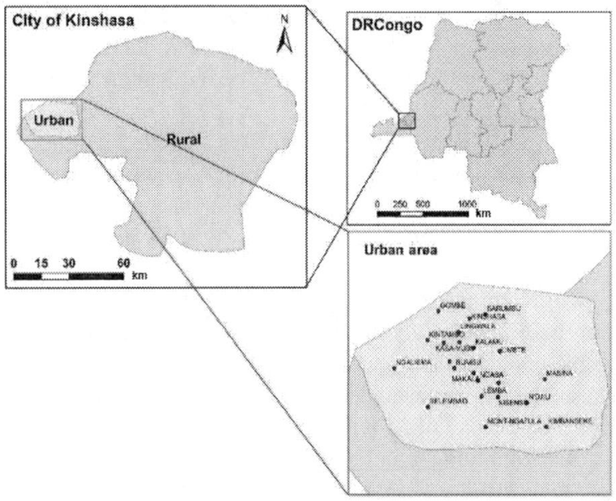

Figure 1. Location of the study area.

it is a hot and humid tropical climate, characterized by a long rainy season lasting up to eight months and often interrupted by short dry periods between January and February (Kayembe and Wolff, 2009). The temperature varies during the year from 20 to 32 °C and infrequently falls under 19 or above 34 °C. Kinshasa soil types constitute Dystric Regosols, Ferralic Arenosols, and Orthic Greyzems. The texture is generally sand, sandy loam, and coarse grained sands (FAO, 1997). The vegetation is generally savanna dotted with shrubs combined with gallery forests at low density areas. Semi deciduous sub-equatorial secondary forests and shrub lands of Guinean are observed in the study area (Kifukieto et al., 2014).

3 Study methodology

3.1 Soil erosion modelling

The study of soil erosion can be traced from centuries, but many approaches that are currently in use began over 70 years ago when Austin Zing published a relationship between soil erosion by water and land slope and length (Ghanshyam, 2012). This was followed by Dwight Smith's relationship that expanded the introduction of conservation practices in the equation (Laflen and Flanagan, 2013). One of the approaches is the Universal Soil Loss Equation (USLE, Laflen and Flanagan, 2013; Kithara et al., 2000), which is a set of mathematical equations that estimate average annual soil loss and sediment yield caused by inter-rill and rill erosion (Toy and Foster, 1998) based on rainfall pattern, soil type, topography, crop system and management practices, and has been widely used USLE (Eq. 1) in the studies of soil erosion management.

$$A = R \cdot K \cdot SL \cdot C \cdot P \qquad (1)$$

Where: A = long term average annual soil loss in tonnes per hectare per year, originally calculated in tonnes per acre per year) ($\mathrm{t\,ha^{-1}\,yr^{-1}}$); R = rainfall-runoff erosivity factor in ($\mathrm{MJ\,mm\,ha^{-1}\,h^{-1}\,yr^{-1}}$); K = Soil erodibility factor ($\mathrm{T\,h\,MJ^{-1}\,mm^{-1}}$); SL = Slope steepness and length factors (dimensionless); C = Cover crop/vegetation management factor (dimensionless), and P = support practice factor which is equal to one (dimensionless) (Laflen and Flanagan, 2013).

While the availability of data required to assess soil erosion based on the above mentioned USLE model is always a challenge for data scarce regions such as the city of Kinshasa, the newly developed Water Observation and Information System (WOIS; TIGER-Net, 2014) provides an advantage of combining remote sensing input with soil erosion factors to determine soil loss potential and erosion hazard.

WOIS is an open source software for assessing water resources using Earth Observation (EO) data. The software was developed by the TIGER initiative under the European Space Agency (ESA) programme aimed at supporting at supporting the African Earth Observation Capacity for Water Resource Management and Monitoring (Guzinski et al., 2014). It uses the Quantum GIS (QGIS) software that acts as the front-end Graphical-User-Interface (GUI) with functionalities that incorporate algorithms for different spatial data analysis libraries.

3.2 Model parameters and data availability

As previously mentioned, while studies on soil erosion are required for a better planning of soil conservation practices in developing countries, it should be stressed that many of these studies have been challenged by the lack of appropriate data over the required spatial and temporal scales. In the present study, global sources of earth observation data are used to obtain the WOIS set of soil erosion parameters.

3.2.1 Rainfall-runoff erosivity factor (*R*-factor)

The amount of rainfall and the peak intensity sustained over an extended period are mainly indicated by the rainfall erosivity. Within WOIS, the R-factor is calculated based on the following relationship:

$$R = 0,\ 4669X - 12,\ 1415 \tag{2}$$

Where: R is the rainfall-runoff erosivity factor expressed in $\mathrm{MJ\,mm\,ha^{-1}\,h^{-1}\,yr^{-1}}$, and X represents the mean annual rainfall in mm. The mean annual rainfall used in this study was calculated from the Climate Hazards Group Infrared Precipitation with Station data (CHIRPS) global monthly average rainfall (1991–2014) with 5 km spatial resolution.

3.2.2 Soil erodibility Factor (*K*-Factor)

K-Factor is the quantitative measure of susceptibility and resistance of soil to erosion. The factor is calculated based on soil texture classes. In this case the soil texture classes from the FAO Digital Soil Map of the World (FAO, 1997) was used to determine K-Factor based on the average values of the erodibility classes from the USLE fact sheet (FAO, 1997).

3.2.3 Slope steepness and length (*SL*-Factor)

The SL factor includes both slope length (Eq. 3) and slope steepness (Eq. 4).

$$L = (X/22,\ 13) \cdot m \tag{3}$$

Where: X is the slope length in meter and m is the weighting factor depending on the slope steepness.

$$S = \left[0.065 + 0.0456(\mathrm{slope}) + 0.006541(\mathrm{slope})^2\right] \tag{4}$$

The SL factor is dimensionless. In this study, a digital elevation model (30 m) was used to determine the SL-factor. The slope classes proposed by Nachtergaele et al. (2010) indicate six categories of slopes ranging from Flat (0–2 %), Undulating (2–8 %), Rolling (8–15 %), moderately steep (15–30 %), Steep (30–60 %), Very steep (> 60 %), and was used for slope classification in this study (Tshimanga, 2012).

3.2.4 Crop/vegetation and management factor (*C*-Factor)

C-Factor is obtained by assigning values of the land cover classes, which were obtained from the Glob Land Cover map with 300 m spatial resolution (CCI-LC-PUG, 2016). The algorithm assigns the C-Factor values to the specific or averaged values related to each land cover type. The table of value is available at European Space Agency, Climate Change Initiative website (http://cci.esa.int/, last access: 23 October 2016).

3.2.5 Support or Erosion Control Practice (*P*-Factor)

P-Factor is the influence of conservation measures on erosion for a particular area. P-Factor is representing the soil loss ratio of the conservation practice to up and down hill culture. The P-Factor ratio is usually less than 1 if the erosion control practice is effective (Gitas et al., 2009; Roger and Herbert, 2016). For this study, P-Factor is considered as 1 because in Kinshasa there is usually no erosion control practices involved.

4 Results and discussion

4.1 Evaluation of the soil erosion parameters

The results of soil erosion parameters for the city of Kinshasa are presented in Table 1.

Rainfall-runoff erosivity (R), the values of R found in this study range from 616 to 723 $\mathrm{MJ\,mm\,ha^{-1}\,h^{-1}\,yr^{-1}}$. Kinshasa

Table 1. Soil erosion parameters.

Soil erosion parameters	Min	Max	SD
Rainfall-runoff erosivity (R), in MJ mm ha^{-1} h^{-1} yr^{-1}	616	723	75
Soil erodibility (K), in Th MJ^{-1} mm^{-1}	0	0.290	0.205
Slope steepness and length (SL),	0	6597	4664
Crop/vegetation and management (C),	0	0.525	0.371

SD – standard deviation.

Table 2. Erosion Hazard and soil loss potential of city of Kinshasa.

Erosion Hazard/Soil loss (t ha^{-1} yr^{-1})	Kinshasa		Rural area		Urban area	
	Area (ha)	%	Area (ha)	%	Area (ha)	%
Very low (< 1)	1 000 280	94.1	918 119	94.2	41 784	92.7
Low (1.1–3)	14 850	1.4	40 234	1.4	324	0.7
Low to medium (3.1–5)	8874	0.8	14 526	0.8	306	0.7
Medium to high (5.1–10)	8977.5	0.8	8568	0.8	454.5	1
High (10.1–15)	4898.25	0.5	8523	0.5	306	0.7
Very high (> 15)	25 013.25	2.3	4590	2.3	1917	4.3

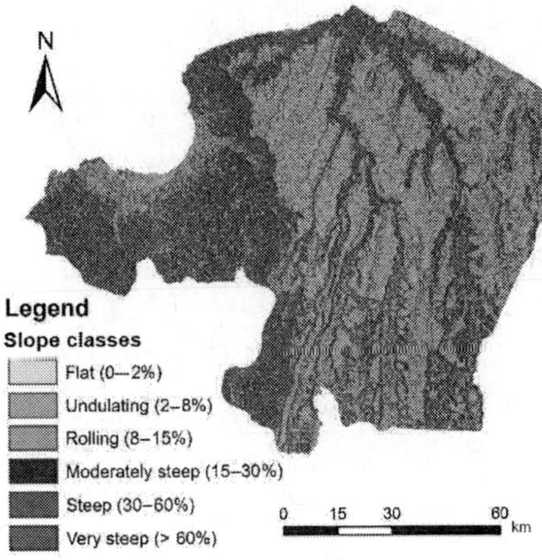

Figure 2. Slope classes of the study area.

receives precipitation characterized by thunderstorms which generate a high volume of rainfall that in turn generate runoff over a long period of time (CPSRP, 2011). When the rainfall occurs with the greater intensity, it exposes the rivers beds to the high risk of transport and sedimentation (Laflen and Flanagan, 2013; Mahabaleshwara and Nagabhushan, 2014). In such process the R-Factor values is high with greater erosivity.

Soil erodibility (K), the values of K found in this study range from 0 to 0.290 Th MJ^{-1} mm^{-1}. K-Factor is the quantitative measure of susceptibility and resistance soil to erosion. The Kinshasa soil characteristics are high percent content of fine sand particles and low organic matter content. The soil are constitute of smaller particles detachable and easily transportable. Such kind of soils are mostly erodible. The positive side of Kinshasa soil is that the soil is permeable which lead to less runoff and reduce the risk of soil erosion at some areas (FAO, 1997).

Slope steepness and length (SL), generally, the results show that Kinshasa does not have a high concentration of steep areas and most of the areas are gentle undulating and rolling. Relatively high SL factor values are found on areas around the rivers (Fig. 2).

Crop/vegetation and management (C), From the Kinshasa Land cover map the C values range from 0 to 0.525. The C-Factor values were assigned to the specific or averaged values related to each land cover type. The table of values is available at the European Space Agency, Climate Change Initiative website. The values can vary from 0 to 1 depending on the land cover type. Thick and dense vegetation offers higher protection against erosion and thus has a lower C-factor values compared to bare soil with no protection against erosion and has a C-factor of 1.

4.2 Soil loss and Erosion hazard estimates

The soil loss using USLE equation is calculated by multiplying all four single factors including R, K, SL, and C. Figures 4 and 5 show the spatial variation of soil loss potential and erosion hazard of the city of Kinshasa. The Table 2 shows the results of soil loss in t ha^{-1} yr^{-1}, it can be realise that these results are over estimate the soil loss when comparing to the results obtained in the previews studies in Kinshasa by Mfumu et al. (2012) where the soil loss vary from 0 to 9 t ha^{-1} yr^{-1}. At national level, the study by Alain

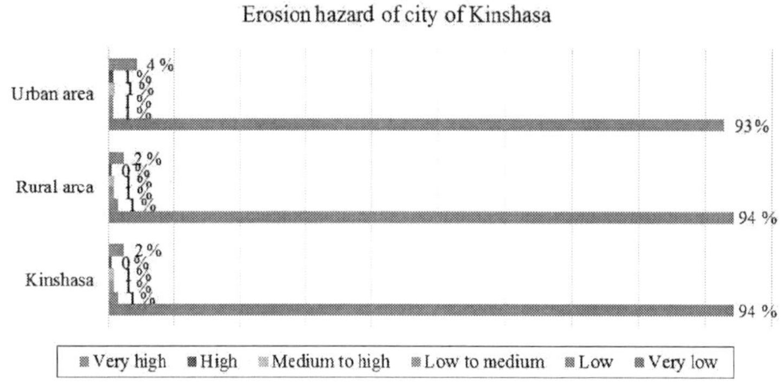

Figure 3. Histogram of erosion hazard.

Figure 4. The spatial variation of soil loss potential.

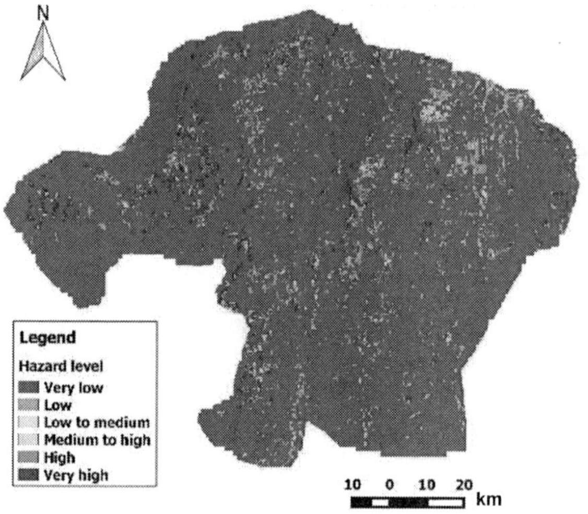

Figure 5. The spatial variation of soil erosion hazard.

et al. (2009) in the Upper Congo, tributaries shows the average rate of 8.9 range $\mathrm{t\,ha^{-1}\,yr^{-1}}$ of soil loss in Kasai and major right-bank tributaries. So the approach used by WOIS to estimate the soil loss is over estimate the soil loss potential at local and national level. At international level in Central Africa particularly Walling and Webb (1996) found the range of 0 to $50\,\mathrm{t\,km^{-2}\,yr^{-1}}$ where the results obtain by WOIS were under estimate. The city of Kinshasa was found with 2.3 % of area with very high risk of erosion. Same situation to the rural part of the city, contrarily to urban part where the area of very high risk of erosion was 4.3 %. The spatial variation maps show that the highest risk of erosion was found along the river or at the area with high elevation and informal settlements. It can be observed that in the urban area the most affected areas were the municipalities of Binza Kinsuka, Mbudi, Malweka, Kimbwala, Binza gendarmerie, Binza pigion, Gombele, Binza IPN, Binza Djelo, Jamaic, Cite mama Mobutu, Cite verte, cite pumbu, Livulu, Mbasa-lemba (Figs. 1 and 5).

5 Conclusion and recommendations

The city of Kinshasa faces challenges of land degradation due to soil erosion, which threaten livelihood of population. This study has shown the strength of GIS based approaches to estimate soil loss potential and erosion hazard. The major conclusions that arise from this study are:

- The urban part of Kinshasa is more prone to soil erosion than the rural counterpart. The high percentage of very high risk of erosion (4.3 %) was found in the urban part when compared to the rural part of the city (2.3 %);
- Comparing to the previous studies at local level the approach used by WOIS over estimates the amount of soil loss. At regional level the results show under estimation;
- The study recommends the soil loss ground measurements in Kinshasa for the validation of the results obtained from the approach used by WOIS;

- The validation will simplify the field work by the use of remote sensing product and approach (WOIS) when doing soil loss and water resources studies.

Data availability. Different sources of data used in this study are:

- Digital Elevation Model (DEM) 30 m of the city of Kinshasa downloaded from https://earthexplorer.usgs.gov/ (last access: 20 April 2018).
- The Glob land Cover map (2008–2012) with 300 m spatial resolution was obtained from the European Space Agency, Climate Change Initiative (CCI-LC-PUG, 2016; http://maps.elie.ucl.ac.be/CCI/viewer/download.php).
- The mean annual rainfall was calculated from the Climate Hazards Group Infrared Precipitation with Station data (CHIRPS) global monthly average rainfall (1991–2014) with 5 km spatial resolution, the CHIRPS rainfall was obtained from the CHIRPS website (ftp://ftp.chg.ucsb.edu/pub/org/chg/products/CHIRPS-2.0/global_2-monthly/tifs/, last access: 20 April 2018).
- The USDA texture classes of the FAO Digital Soil Map of the World (FAO, 1997). And K-Factor for different soil types are available at http://www.omafra.gov.on.ca/english/engineer/facts/12-051.htm (last access: 20 April 2018).

Competing interests. The authors declare that they have no conflict of interest.

Special issue statement. This article is part of the special issue "Understanding spatio-temporal variability of water resources and the implications for IWRM in semi-arid eastern and southern Africa". It is a result of the IAHS Scientific Assembly 2017, Port Elizabeth, South Africa, 10–14 July 2017.

Acknowledgements. We acknowledge WATERNET/SADC for supporting our participation to the IAHS Scientific Assembly 2017, Elizabeth, South Africa, 10–14 July 2017.

Edited by: Hodson Makurira
Reviewed by: Timothy Dube and one anonymous referee

References

Alain, L., Bricquet, J. P., Pandi, A., and Olivry, J. C.: A review of material transport by the Congo River and its tributaries, Hydrol. Process. 23, 3216–3224, https://doi.org/10.1002/hyp.7395, 2009.

Antoine, B.: Les concepts de la géographie humaine, 2nd Edn., Masson, Paris, 1991.

Bhattacharyya, R., Birendra, N. G., Prasanta, K. M., Biswapati, M., Cherukumalli, S. R., Dibyendu, S., Das, K.: Soil Degradation in India: Challenges and Potential Solutions, Sustainability, 7, 3528–3570, https://doi.org/10.3390/su7043528, 2015.

Bizuwerk, A., Taddese, G., and Getahum, Y.: Application of GIS for Modeling Soil Loss Rate in Awash River Basin, Ethiopia, Addis Ababa, available at: https://www.academia.edu/8344707/Application_of_GIS_for_Modeling_Soil_loss_rate_in_Awash_River_Basin_Ethiopia (last access: 20 April 2018), 2004.

CCI-LC-PUG: Land Cover Climate Change Initiative Product User Guide Version 2, Bruxelles, available at: https://maps.elie.ucl.ac.be/CCI/viewer/download/ESACCI-LC-Ph2-PUGv2_2.0.pdf (last access: 20 April 2018), 2016.

Daniel, J., Karl, W. J., Pamela, J., and Jon, E.: Soil Erosion in Humid Regions: A Review, Journal of Contemporary Water Reseach and Education, 154, 48–59, 2015.

FAO: Soil Map of the World, Rome: World Soil Information(ISRIC), available at: http://www.fao.org/fileadmin/user_upload/soils/docs/isricu_i9264_001.pdf (last access: 21 March 2017), 1997.

Ghanshyam, D.: Hydrology and soil conservation engineering: including watershed management, 2nd Edn., PHI Learning Pvt. Ltd., New Delhi, 2012.

Gitas, I. Z., Douros, K., Minakou, C., and Silleos, G. N.: Multi-Temporal Soil Erosion Risk Assessment in N. Chalkidiki Using a Modified USLE Raster Model, Earsel eProceedings, 8, 40–52, 2009.

Guzinski, R., Kass, S., Silvia, H., Bauer-gottwein, P., Hedegaard, I. J., Vahid, N., Doubkova, M., Walli, A., and Tottrup, C.: Enabling the Use of Earth Observation Data for Integrated Water Resource Management in Africa with the Water Observation and Information System, Remote Sens., 6, 7819–7839, https://doi.org/10.3390/rs6087819, 2014.

Kayembe, M. M. and Wolff, E.: Cartographie de La Croissance Urbaine de Kinshasa (R.D. Congo) Entre 1995 et 2005 Par Télédétection Satellitaire À Haute Résolution, Belgeo, 4, 439–455, https://doi.org/10.4000/belgeo.7349, 2009.

Kifukieto, C., Deligne, J., Kachaka, C., and Francis, F.: Inventaire Préliminaire Des Termites (Isoptera) Du Plateau Des Batékés En République Démocratique Du Congo, Entomologie Faunistique, 67, 49–57, 2014.

Kinshasa-CPSRP: Monographie de La Ville de Kinshasa, Vol. 14754, Kinshasa, 2011.

Kitambo, B. M.: Estimation of erosion and sediment yield using EPM model in the ungauged N'sele River catchment, Democratic Republic of Congo, University of Dar es Salaam, MSC thesis, University of Dar es Salaam, Tanzania, 2017.

Kithara, H., Okura, Y., Sammori, T., and Kawanami, A.: Application of Universal Soil Loss Equation (USLE) to Mountainous Forests in Japan, J. For. Res.-Jpn., 5, 231–236, https://doi.org/10.1007/BF02767115, 2000.

Laflen, J. M. and Flanagan, D. C.: The Development of U.S. Soil Erosion Prediction and Modeling, International Soil and Water Conservation Research, 1, Elsevier Masson SAS, 1–11, https://doi.org/10.1016/S2095-6339(15)30034-4, 2013.

Mahabaleshwara, H. and Nagabhushan, H. M.: A Study on Soil Erosion and Its Impacts on Floods and Sedimentation, International Journal of Research in Engineering and Technology, 3, 443–51, 2014.

Mfumu, K. A., Ndembo, L. J., Lukanda, M. V., Nlandu, W. J., Kombele, D. G., and Bola, B. G.: Analyse spatiale du risque d'erosion sur le bassin versant de la riviere kemi et impacts potentiel sur la surete des installations nucleaires de Kinshasa", Revue Congolaise des Sciences Nucléaires, Volume 26, No. 2, 2012.

Nachtergaele, A. F., van Velthuizen, B. H., Verelst, B. L., Batjes, C. N., Dijkshoorn, C. K., van Engelen, C. V., Fischer, B. G., Jones, D. A., Montanarella, D. L., Petri, A. M., Prieler, B. S., Shi, E. X., Teixeira, D. E., and Wiberg, D. D.: World Congress of Soil Science, Soil Solutions for a Changing World, 1–6 August 2010, Brisbane, Australia, 2010.

Pimentel, D. and Burgess, M.: Soil Erosion Threatens Food Production, Agriculture, 3, 443–463, https://doi.org/10.3390/agriculture3030443, 2013.

Robertson, S. A.: Access to Water in the Slums of the Developing World, International Policy Centre for Inclusive Growth Working Papers, no. 57, International Policy Centre for Inclusive Growth, Brasilia, 2009.

Roger, J. and Herbert, B.: Application of the USLE to Southwestern Rangelands, Hydrology and Water Resources, 2016.

Telles, T. S., Carmela, S., Dechen, F., Gustavo, L., De Souza, A., and De Fátima Guimarães, M.: Valuation and Assessment of Soil Erosion Costs, Scientia Agricola, 70, 209–216, 2013.

TIGER-Net: Water Observation and Information System and Product Portfolio, TIGER-Net Project, Bruxelles, available at: www.tiger-net.org (last access: 12 December 2017), 2014.

Toy, T. and Foster, G.: Use of the Revised Universal Soil Loss Equation (RUSLE) on Mined Lands, Construction Sites and Reclaimed Lands, Denver, available at: https://www.osmre.gov/resources/library/ghm/rusleGuidelines.pdf (last access: 20 April 2018), 1998.

Tshimanga, R. M.: Hydrological Uncertainty Analysis and Scenario-Based 83 Streamflow Modelling for the Congo River Basin, Rhodes University, Grahamstown, South Africa, 2012.

Walling, D. E. and Webb, B. W.: Erosion and sediment yield: a global overview over the world particularly for almost the central part of Africa, Proceedings of the Exeter Symposium, July 1996, IAHSPubl. no. 236, 1996.

https://doi.org/10.5194/piahs-378-59-2018

Open Access
PIAHS

Evaluation of sub daily satellite rainfall estimates through flash flood modelling in the Lower Middle Zambezi Basin

Thomas Matingo, Webster Gumindoga, and Hodson Makurira

Department of Civil Engineering, University of Zimbabwe, P.O. Box MP167, Mt Pleasant, Harare, Zimbabwe

Correspondence: Thomas Matingo (engineermatingo@gmail.com), Webster Gumindoga (wgumindoga@gmail.com), and Hodson Makurira (hmakurira@yahoo.com)

Received: 12 November 2017 – Revised: 12 January 2018 – Accepted: 15 January 2018 – Published: 29 May 2018

Abstract. Flash floods are experienced almost annually in the ungauged Mbire District of the Middle Zambezi Basin. Studies related to hydrological modelling (rainfall-runoff) and flood forecasting require major inputs such as precipitation which, due to shortage of observed data, are increasingly using indirect methods for estimating precipitation. This study therefore evaluated performance of CMORPH and TRMM satellite rainfall estimates (SREs) for 30 min, 1 h, 3 h and daily intensities through hydrologic and flash flood modelling in the Lower Middle Zambezi Basin for the period 2013–2016. On a daily timestep, uncorrected CMORPH and TRMM show Probability of Detection (POD) of 61 and 59 %, respectively, when compared to rain gauge observations. The best performance using Correlation Coefficient (CC) was 70 and 60 % on daily timesteps for CMORPH and TRMM, respectively. The best RMSE for CMORPH was 0.81 % for 30 min timestep and for TRMM was 2, 11 % on 3 h timestep. For the year 2014 to 2015, the HEC-HMS (Hydrological Engineering Centre-Hydrological Modelling System) daily model calibration Nash Sutcliffe efficiency (NSE) for Musengezi sub catchment was 59 % whilst for Angwa it was 55 %. Angwa sub-catchment daily NSE results for the period 2015–2016 was 61 %. HEC-RAS flash flood modeling at 100, 50 and 25 year return periods for Angwa sub catchment, inundated 811 and 867 ha for TRMM rainfall simulated discharge at 3 h and daily timesteps, respectively. For CMORPH generated rainfall, the inundation was 818, 876, 890 and 891 ha at daily, 3 h, 1 h and 30 min timesteps. The 30 min time step for CMORPH effectively captures flash floods with the measure of agreement between simulated flood extent and ground control points of 69 %. For TRMM, the 3 h timestep effectively captures flash floods with coefficient of 67 %. The study therefore concludes that satellite products are most effective in capturing localized hydrological processes such as flash floods for sub-daily rainfall, because of improved spatial and temporal resolution.

1 Introduction

Accurate measurement of precipitation is key for dependable hydrologic predictions (Guo et al., 2015). Rain gauges are traditionally used to measure precipitation and have a drawback of reduced spatial coverage (Blacutt et al., 2015) thus rendering hydrological predictions inaccurate (Ning et al., 2016). it is impossible to place rain gauge stations at all places of interest. Therefore, satellite rainfall helps to augment the sparse rain gauges coverage (Maidment Ross et al., unpublished data, 2017). Nowadays satellite rainfall products such as Climate Prediction Centre Morphing technique (CMORPH), Tropical Rainfall Measuring Mission (TRMM) and Precipitation Estimation from Remotely Sensed Information using Artificial Neural Networks (PERSIANN) are produced at a spatial resolution of 8 km and temporal resolution of 3 h. Due to their indirect nature of measurements, satellite rainfall estimates (SREs) contain errors which can be divided into two groups; (i) systematic (bias) and (ii) random errors (Liu, 2015). As such, bias correction is needed before SREs are used with confidence in water resources applications. Previous studies in the Middle Zambezi Basin such as by Samuel Mwanangombe (unpublished data, 2015) only concentrated on highlighting the spatial aspects in SREs but ignored the temporal aspect in the hydrological outputs. Flash floods occur within a short period of time such that daily time scale used in previous studies may not be adequate

Published by Copernicus Publications on behalf of the International Association of Hydrological Sciences.

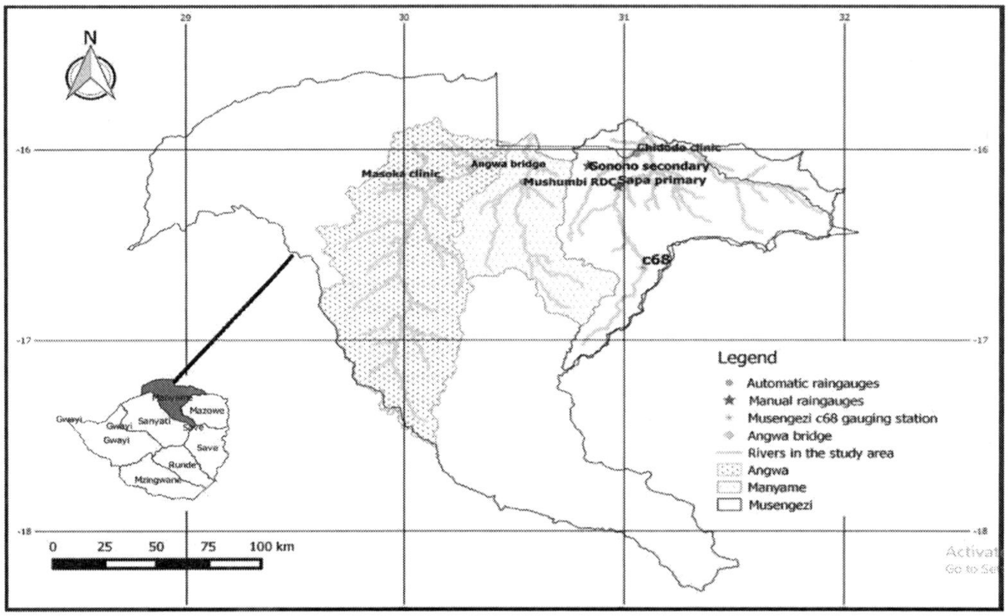

Figure 1. The Mbire District in Lower Guruve showing manual and automatic rain gauge stations.

to capture flash floods (Saber and Yilmaz, 2016). Floods are mostly localized. Besides the need for high spatial resolution to focus on floods, there is also need for temporal resolution such as sub daily time scales to capture the actual flood processes (Chen et al., 2016). By their improved spatial coverage and ability to provide rainfall estimates at recurrent time steps. SREs are suitable for basin-wide applications such capturing flash floods in cases where rain gauge measurements are not available. This is a welcome development in study sites such as Mbire District where there is limited spatial and temporal cover of automated recording rain gauges. The objective of this study was therefore to assess the performance of SREs at finer temporal resolutions and then applying the generated data in rainfall-runoff and flash flood modelling in the middle Zambezi Basin.

2 Study Area

The study area covers 3 sub-catchments of Manyame cutting through the Mbire District of Zimbabwe, namely, Angwa, Lower Manyame and Musengezi. Mbire District lies between 30.60 and 31.20° E and 15.60 and 16.40° S. The estimated area of Mbire District is approximately 4700 km^2 and is characterized mainly by the floodplains of the Zambezi River Basin. The altitude is approximately 400 m above sea level and being drained by the Angwa, Manyame, Musengezi and Kadzi rivers (Fig. 1). Mbire District receives low rainfall of below 600 mm and lies in agro ecological region IV of Zimbabwe (Bola et al., 2014). The district records high temperatures (sometimes exceeding 40 °C) with mean annual temperature of 25 °C, Mbire District lies on sedimentary geological foundations of lime and sandstone formations. Varied soils rich in sodium exist though lacking in organic matter. Mbire District experiences frequent localised flooding (CPU, unpublished data, 2015). Backflow from the downstream Cahora Bassa Dam and rainfall events from the large upstream catchment are the two main factors that contribute to flooding (Bola et al., 2014). Angwa, Manyame, Musengezi sub-basins have been selected since these are the major rivers in the area and greatly contribute to flooding (Bola et al., 2014).

3 Methodology

3.1 Gauge based rainfall data

Rainfall for November 2015 to December 2016 with temporal resolution of 30 min was acquired from three automatic weather stations located at Chidodo, Masoka and Mushumbi. The 30 min data was aggregated to 1 h, 3 h and daily timestep. Before 2015, there was no sub-daily rainfall data available from rain gauges. The automatic rain gauges were installed in October 2015 and these provide sub daily temporal resolution rainfall data. One year of observations would not lead to convincing conclusions hence the need to lengthen the period of analysis using the "Hydrognomon" software (Kozanis et al., 2005). The daily time series for 2013 to October 2015 was disaggregated to lengthen the comparison time series.

3.2 Satellite rainfall download and processing

TRMM and CMORPH rainfall data were downloaded using ILWIS 3.7.2 software through the GeoNETCAST ISOD toolbox. The point to pixel method was used for comparison of gauge rainfall and satellite rainfall. A maplist was created and crossed to the rain gauge stations point map to extract satellite rainfall values.

The rain gauge stations point map was used to extract satellite rainfall values from the grid CMORPH and TRMM rainfall.

3.3 Performance evaluation of satellite products through standard statistics

CMORPH and TRMM Satellite rainfall for 2013 to 2015 were compared with rain gauge data by means of detection statistics and standard statistical analysis. The detection statistics and statistical analysis used are POD, RMSE, CC and Rbias using the following equations:

$$\mathrm{POD} = H/H + M \tag{1}$$

$$\mathrm{FAR} = F/H + F \tag{2}$$

$$\mathrm{FBIAS} = F + H/H + M \tag{3}$$

$$\mathrm{CSI} = H/H + M + F \tag{4}$$

$$\mathrm{Rbias} = \frac{\varepsilon(P_\mathrm{s} - P_\mathrm{r})}{\varepsilon P_\mathrm{r}} \tag{5}$$

$$\mathrm{RMSE} = \sqrt{\frac{(P_\mathrm{s} - P_\mathrm{r})^2}{N}} \tag{6}$$

$$\mathrm{CC} = \frac{\varepsilon(P_\mathrm{r} - P_\mathrm{mr})(P_\mathrm{s} - P_\mathrm{ms})}{\sqrt{\left(\varepsilon(P_\mathrm{r} - P_\mathrm{mr})^2\right)}\sqrt{\varepsilon(P_\mathrm{s} - P_\mathrm{ms})^2}} \tag{7}$$

Where H = hit bias, M = miss bias, F = false bias, FAR = false alarm ratio, POD = probability of detection, CSI = critical success index, FBIAS = frequency bias, P_s = satellite rainfall estimates (mm day^{-1}), P_ms = mean values of SRE (mm day^{-1}), P_r = rain gauge rainfall (mm day^{-1}), P_mr = mean values of rain gauge rainfall (mm day^{-1}), N = sample size in days.

3.4 Bias correction of satellite rainfall

For correcting the satellite rainfall, a correction factor ($\mathrm{BF_{stb}}$) was calculated using Eq. (8). The multiplicative bias factor corrects for the spatial and temporal aspects of the error in satellite rainfall (Bhatti et al., 2016; Habib et al., 2014; Gumindoga et al., 2016a).

$$\mathrm{BF_{stb}} = \frac{\sum_{t=d}^{t=d-l} S(i,t)}{\sum_{t=d}^{t=d-l} G(i,t)} \tag{8}$$

Where G and S = daily gauge and satellite rainfall estimates, respectively (mm), i = gauge location, t = julian day number (day), l = length of a time window for bias calculation (days), n = the total number of gauges within the entire domain of the study, T = full duration of the study period.

3.5 HEC HMS model setup

3.5.1 Land surface parameterization

A Shuttle Radar Topography Mission (SRTM) Digital Elevation Model (DEM) at 90 m resolution was obtained from the following website: http://srtm.csi.cgiar.org/. The DEM was imported in ILWIS for DEM hydro processing. Catchment merging was conducted to derive the 3 sub-catchments. The catchment maps were exported from ILWIS as ESRI shapefiles together with the river shapefiles after which a basin model was created in HEC-HMS (Abushandi and Merkel, 2013; Gumindoga et al., 2016b; Narayan and Gautam, 2015). The deficit and constant loss method and the SCS unit Hydrograph Transformation methods were used for the component (Bhuiyan et al., 2017). The Muskingum method was used for reach routing.

3.5.2 Model calibration, validation and model efficiency assessment

The HEC HMS 3.5 model was manually calibrated for daily timestep simulation with Angwa and Musengezi catchments flow dataset for the period 1 October 2014 to 30 September 2015. The calibrated parameters were time lag and percentage impervious (Gumindoga et al., 2016b). Simulated runoff was compared with the measured runoff for Angwa and Musengezi catchments. The performance indicators used to assess the model differences between observed and simulated were the Nash-Sutcliffe efficiency (NSE) and Relative Volume Error (RVE). For validation of the model, one hydrological year; October 2015 to September 2016 was used. All the parameters obtained during model calibration were maintained but the hydro-meteorological series were changed.

3.5.3 Runoff simulation for different timesteps

The loss and transformation parameters for the calibrated daily timestep were transferred to 30 min, 1 h and 3 h timesteps. To check the performance for these timesteps; total volume, time of peak and peak discharge were compared to the standard daily time step gauge.

3.6 HECRAS flash flood modelling

The Angwa sub-catchment 30 m SRTM DEM was used to create a TIN in HEC-GeoRAS. Geometric data relating to centreline, cross section lines, river banks and flow path from ArcView was exported to HECRAS to generate water surface profiles and flood way boundaries. Discharge for different timesteps from HEC-HMS was used to evaluate the performance. For each discharge, standard deviation and the mean were calculated using the Gumbel method for 100, 50 and 25 year flood events. The measure of agreement was calculated using Cohen Kappa in SPSS package for the validation of the flood simulation.

4 Results and Discussion

4.1 Evaluation of satellite rainfall product

Figure 2 shows standard statistics error at different timesteps. The Correlation Coefficient (CC) for TRMM and CMORPH decreases with decreasing timestep. Shen et al. (2010) evaluated performance of TRMM and CMORPH in China at different timescales such as daily, 12, 6, 3 h and found out that CC increases with increase in timestep which is similar to the finding of this study. POD for all satellite products decreases with decrease in timestep such as from daily to 3 h. CMORPH decreases from 0.61 to 0.5 whilst TRMM from 0.59 to 0.31. FBI was smaller than one for TRMM satellite product meaning that it have a tendency of underestimating rainfall whilst CMORPH; FBI was 2.02 which shows a tendency of overestimation which is in agreement with results by Yang and Luo (2014) who found FBI of 4.6 and also a study in the Zambezi Basin by Cohen Liechti et al. (2012) who found FBI of 2.9 at daily timesteps.

4.2 Bias correction results

After bias correction, the RMSE reduced by 12 % and Rbias by 33 % on daily time step. This finding of reduction of RMSE after bias correction application tallies with the findings by Gumindoga et al. (2016a), who reduced RMSE by 11 % with similar bias correction on a daily timestep in the Zambezi River Basin. The effect of reducing RMSE after application of bias correction was also found by Bhatti et al. (2016) who decreased RMSE by 7 % on daily timestep in the Gilgel Abbey catchment.

4.3 HEC-HMS calibration and validation with rain gauge rainfall

The calibration results for Angwa and Musengezi upstream for the period of October 2014 to September 2015 are shown in Fig. 3. The Nash Sutcliffe (NSE) was 55 % and Relative Volume Error (RVE) was −9.4 % which is within acceptable range of −10 and 10 % (Janssen and Heuberger, 1995). The negative sign shows that the model underestimated runoff. This can be attributed to the assumption made that ignored the contribution of base flow in the calibration of the model. For Musengezi upstream, the NSE was 58 % which shows that the model reproduced the observed patterns fairly well. The relative volume error was −7.7 %. The RVE differs from Gumindoga et al. (2016b) in the same basin which was −1.54 %.

Model validation for the period of October 2015 to September 2016 was carried out successfully for Angwa River with a NSE of 61 % which is within the accepted range. However, the RVE obtained of 42 % is outside the acceptable range. Figure 4 shows the hydrograph for validation.

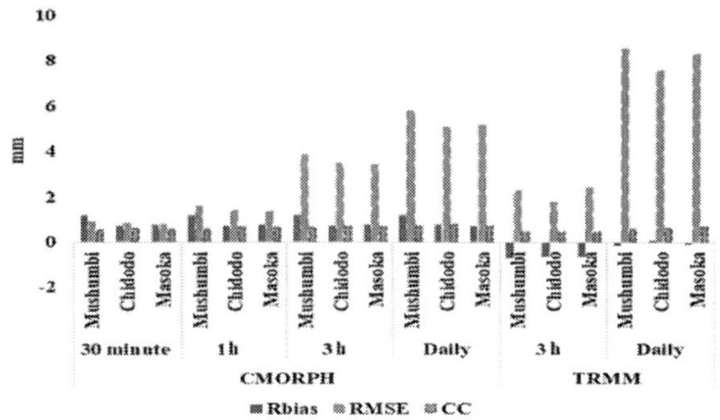

Figure 2. Standard statistic error for both SREs at different timesteps.

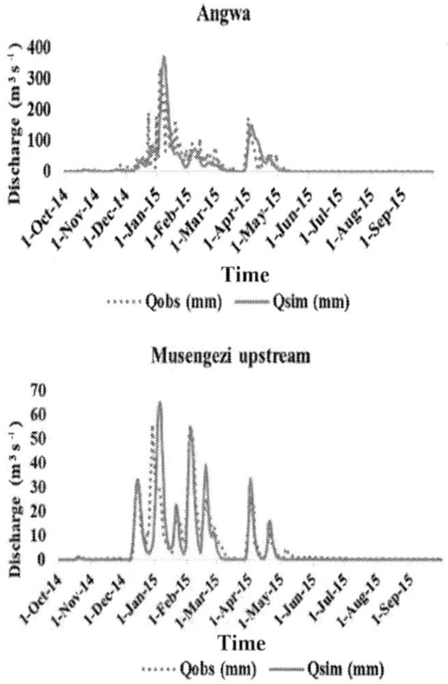

Figure 3. Calibration results for Angwa and Musengezi rivers upstream based on daily timestep.

4.4 HEC-HMS model simulation with SREs at different timesteps

Table 1 was acquired as an output from the HEC-HMS model and it shows that peak discharge increases with decrease in timestep. This finding is consistent with results from Bruce Mcenroe (unpublished data, 2010) which showed that simulated runoff increases with a decrease in timesteps considered. This shows that coarser timesteps do not necessarily imply higher peaks. A finer timestep might record some of the maximum rainfall amounts that might actually be missed by a coarser timestep.

Table 1. Characteristics of catchments acquired from the model basing on different timestep for both satellite rainfall products for October 2014 to September 2015.

Catchments	Drainage area (km^2)	Peak Discharge ($m^3\,s^{-1}$)	Time of peak	Volume ($1000\,m^3$)
Rain gauge rainfall daily data				
Angwa	9502	259.1	7 January 2015	561 686.4
Musengezi upstream	951	38.1	4 January 2015	102 415.9
CMORPH satellite rainfall 3 h data				
Angwa	9502	3135.6	5 January2015	2 898 026.9
Musengezi upstream	951	250.6	2 February 2015	365 226.7
TRMM satellite rainfall 3 h data				
Angwa	9502	106	12 January 2015	285 114.6
Musengezi upstream	951	51.4	5 January 2015	45 950.3
CMORPH satellite rainfall 1 h data				
Angwa	9502	3581.6	5 January 2015	3 185 872.3
Musengezi-upstream	951	276.8	2 February 2015	393 687.5
CMORPH satellite rainfall 30 min data				
Angwa	9502	3689.8	5 January 2015	3 458 287.1
Musengezi-upstream	951	279.6	02 February 2015	420 255.6

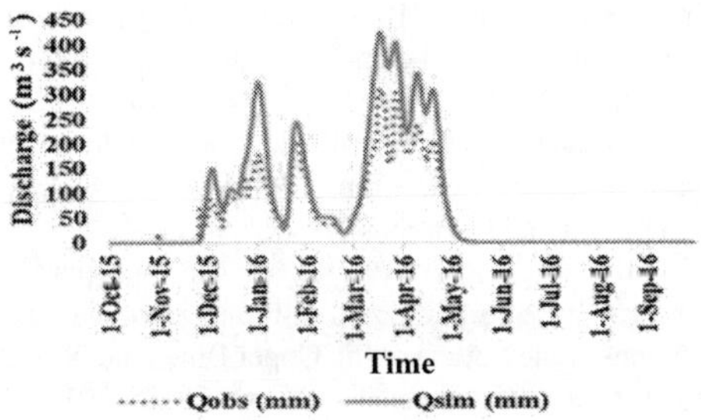

Figure 4. Validation hydrograph for Angwa.

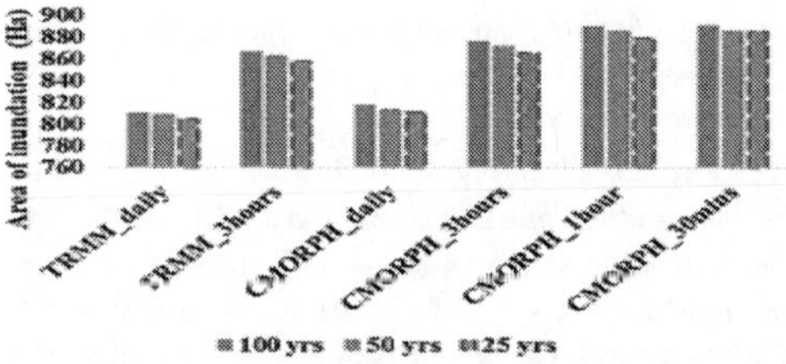

Figure 5. Comparison of area of inundation for two satellites on different time step.

4.5 Application of bias corrected SREs into flash flood modelling

Figure 5 shows the area of inundation. For both satellites, flood extent for daily timestep was smaller than 3 h timestep. This confirms that flash floods, by their name, occur rapidly within a short period of time and immediately disappear such that daily timestep captures a lesser extent of flooded area. At a finer timestep such as 30 min and 1 h, CMORPH simulates the flash floods better than TRMM. The area of inundation by TRMM generated discharge is less than that of CMORPH generated discharge.

4.6 Validation of HECRAS flood simulation

The Cohen Kappa statistic shows that the ground control points were in agreement with the simulated flood plain by HECRAS. The coefficients were as follows: ($P < 0.05$) of 69, 67, 66.7, 54, 67 and 62 % for CMORPH 30 min, 1 h, 3 h, daily, TRMM 3 h and TRMM daily satellite precipitation respectively. Finer timesteps capture all floods since they are measured instantaneously whereas coarser timesteps present a mean over a period in question.

Figure 6 shows the best flood extent polygon at 30 min timestep and 100 year return period overlaid on google map. It can be seen that cultivating area is mostly at great risk when floods occur.

5 Conclusions

Both CMORPH and TRMM applications exhibit bias factors which were too large and, hence, warranted bias correction before application in hydrological and hydraulic modelling.

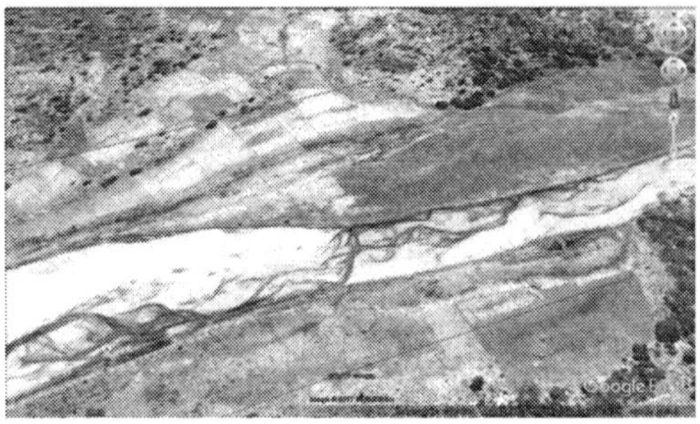

Figure 6. Flood extent polygon for 100 year return period at 30 min timestep at Masoka.

CMORPH also performed better than TRMM in terms of NSE and RVE when applied to the HEC-HMS model.

Since the flooded area was at finer timescales than for the aggregated daily timescales, it can also be concluded that floods occur rapidly and the chances of capturing them are higher when finer resolution of measurement are applied.

Data availability. The data is accessible on request from the authors.

Competing interests. The authors declare that they have no conflict of interest.

Special issue statement. This article is part of the special issue "Understanding spatio-temporal variability of water resources and the implications for IWRM in semi-arid eastern and southern Africa". It is a result of the IAHS Scientific Assembly 2017, Port Elizabeth, South Africa, 10–14 July 2017.

Acknowledgements. WaterNet is highly acknowledged for supporting this research. The authors greatly appreciate the Department of Civil Engineering for providing the platform to carry out the research and helping to shape its successful conclusion.

Edited by: Evison Kapangaziwiri
Reviewed by: Amidu Owolabi Ayeni and one anonymous referee

References

Abushandi, E. and Merkel, B.: Modelling Rainfall Runoff Relations Using HEC-HMS and IHACRES for a Single Rain Event in an Arid Region of Jordan, Water Resour. Manag., 27, 2391–2409, https://doi.org/10.1007/s11269-013-0293-4, 2013.

Bhatti, H. A., Rientjes, T., Haile, A. T., Habib, E., and Verhoef, W.: Evaluation of Bias Correction Method for, Sensors, 16, 10725–10748, https://doi.org/10.3390/s16060884, 2016.

Bhuiyan, H. A. K. M., Mcnairn, H., Powers, J., and Merzouki, A.: Application of HEC-HMS in a Cold Region Watershed and Use of RADARSAT-2 Soil Moisture in Initializing the Model, Hydrology, 4, 1–19, https://doi.org/10.3390/hydrology4010009, 2017.

Blacutt, L. A., Herdies, D. L., Gustavo, L., Gonçalves, G. De, Vila, D. A., and Andrade, M.: Precipitation comparison for the CFSR, MERRA, TRMM3B42 and Combined Scheme datasets in Bolivia, Atmos. Res., 163, 117–131, https://doi.org/10.1016/j.atmosres.2015.02.002, 2015.

Bola, G., Mabiza, C., Goldin, J., Kujinga, K., Nhapi, I., Makurira, H., and Mashauri, D.: Coping with droughts and floods: A Case study of Kanyemba, Mbire District, Zimbabwe, Phys. Chem. Earth, 67–69, 180–186, https://doi.org/10.1016/j.pce.2013.09.019, 2014.

Chen, X., Zhang, L., Gippel, C. J., Shan, L., Chen, S., and Yang, W.: Uncertainty of Flood Forecasting Based on Radar Rainfall Data Assimilation, Hindawi Publ. Corp., Adam House, London, UK, 2016.

Cohen Liechti, T., Matos, J. P., Boillat, J.-L., and Schleiss, A. J.: Comparison and evaluation of satellite derived precipitation products for hydrological modeling of the Zambezi River Basin, Hydrol. Earth Syst. Sci., 16, 489–500, https://doi.org/10.5194/hess-16-489-2012, 2012.

Gumindoga, W., Rientjes, T. H. M., Haile, A. T., Makurira, H., and Reggiani, P.: Bias correction schemes for CMORPH satellite rainfall estimates in the Zambezi River Basin, Hydrol. Earth Syst. Sci. Discuss., https://doi.org/10.5194/hess-2016-33, 2016a.

Gumindoga, W., Rwasoka, D. T., Nhapi, I., and Dube, T.: Ungauged runoff simulation in Upper Manyame Catchment, Zimbabwe: Application of the HEC-HMS model, Phys. Chem. Earth, 5, 1–12, https://doi.org/10.1016/j.pce.2016.05.002, 2016b.

Guo, H., Chen, S., Bao, A., Hu, J., and Yang, B.: Comprehensive Evaluation of High-Resolution Satellite-Based Precipitation Products over China, Atmosphere (Basel), 7, 1–9, https://doi.org/10.3390/atmos7010006, 2015.

Habib, E., Haile, A. T., Sazib, N., Zhang, Y., and Rientjes, T.: Effect of Bias Correction of Satellite-Rainfall Estimates on Runoff Simulations at the Source of the Upper Blue Nile, Remote Sens., 6688–6708, https://doi.org/10.3390/rs6076688, 2014.

Janssen, P. H. M. and Heuberger, P. S. C.: Calibration of process-oriented models, Ecol. Model., 83, 55–66, https://doi.org/10.1016/0304-3800(95)00084-9, 1995.

Kozanis, S., Christofides, A., Mamassis, N., and Efstratiadis, A. K.: Hydrognomon – A hydrological data management and processing software tool, Eur. Geosci. Union, 7, 04644, https://doi.org/10.13140/RG.2.2.34222.10561, 2005.

Liu, Z.: Evaluation of Precipitation Climatology Derived from TRMM Multi-Satellite Precipitation Analysis (TMPA) Monthly Product over Land with Two Gauge-Based Products, Climate, 3, 964–982, https://doi.org/10.3390/cli3040964, 2015.

Narayan, E. and Gautam, P.: Hydrological Modeling with HEC-HMS in Different Channel Sections in Case of Gandaki River Basin, Global Journals Inc, Wakefield, USA, 2015.

Ning, S., Wang, J., Jin, J., and Ishidaira, H.: Assessment of the Latest GPM-Era High-Resolution Satellite Precipitation Products by Comparison with Observation Gauge Data over the Chinese Mainland, Water, 41–42, https://doi.org/10.3390/w8110481, 2016.

Saber, M. and Yilmaz, K.: Bias Correction of Satellite-Based Rainfall Estimates for Modeling Flash Floods in Semi-Arid regions:

Application to Karpuz River, Turkey, Nat. Hazards Earth Syst. Sci. Discuss., https://doi.org/10.5194/nhess-2016-339, 2016.

Shen, Y., Xiong, A., Wang, Y., and Xie, P.: Performance of high-resolution satellite precipitation products over China, J. Geophys. Res., 115, D02114, https://doi.org/10.1029/2009JD012097, 2010.

Yang, Y. and Luo, Y.: Evaluating the performance of remote sensing precipitation products CMORPH, PERSIANN, and TMPA, in the arid region of northwest China, Theor. Appl. Clim., 118, 429–445, https://doi.org/10.1007/s00704-013-1072-0, 2014.

https://doi.org/10.5194/piahs-378-67-2018

Open Access
PIAHS

Optimal water resource allocation modelling in the Lowveld of Zimbabwe

Delight Mhiribidi[1], Joel Nobert[1], Webster Gumindoga[2], and Donald T. Rwasoka[3]

[1]Department of Water Resources Engineering, College of Engineering and Technology, University of Dar es Salaam, Dar es Salaam, Tanzania
[2]Department of Civil Engineering, University of Zimbabwe, P.O. Box MP167, Mt Pleasant, Harare, Zimbabwe
[3]Upper Manyame Subcatchment Council, Box 1892, Harare, Zimbabwe

Correspondence: Delight Mhiribidi (delightmhiribidi@gmail.com)

Received: 6 November 2017 – Accepted: 4 December 2017 – Published: 29 May 2018

Abstract. The management and allocation of water from multi-reservoir systems is complex and thus requires dynamic modelling systems to achieve optimality. A multi-reservoir system in the Southern Lowveld of Zimbabwe is used for irrigation of sugarcane estates that produce sugar for both local and export consumption. The system is burdened with water allocation problems, made worse by decommissioning of dams. Thus the aim of this research was to develop an operating policy model for the Lowveld multi-reservoir system.

The Mann Kendall Trend and Wilcoxon Signed-Rank tests were used to assess the variability of historic monthly rainfall and dam inflows for the period 1899–2015. The WEAP model was set up to evaluate the water allocation system of the catchment and come-up with a reference scenario for the 2015/2016 hydrologic year. Stochastic Dynamic Programming approach was used for optimisation of the multi-reservoirs releases.

Results showed no significant trend in the rainfall but a significantly decreasing trend in inflows ($p < 0.05$). The water allocation model (WEAP) showed significant deficits ($\sim 40\,\%$) in irrigation water allocation in the reference scenario. The optimal rule curves for all the twelve months for each reservoir were obtained and considered to be a proper guideline for solving multi- reservoir management problems within the catchment. The rule curves are effective tools in guiding decision makers in the release of water without emptying the reservoirs but at the same time satisfying the demands based on the inflow, initial storage and end of month storage.

1 Introduction

Effective water resources management requires optimal water resources allocation systems that maximizes set objectives such as crop yield (Reddy and Kumar, 2007); hydroelectric power generation (Mujumdar and Nirmala, 2007) and other uses such as flood control (Sharma et al., 2016). Optimal water allocation is the consideration of all the water users and allocation of water equitably.

Man-made reservoirs are among the most efficient infrastructures to manage water supply as well as minimise the enormous impacts of droughts. These infrastructures, coupled with the need for improved operational skills of watershed-scale reservoirs and water distribution networks necessitates the development and application of water allocation models (Namchaiswadwong, 2004; Sert et al., 2007; Juízo and Lidén, 2010). Simulation-based allocation models are used to meet this need and they simulate water allocation with pre-determined water use priorities and reservoir operation rules. These models have poor flexibility if anthropogenic and hydro-climatic factors are taken into consideration as they operate in a "what-if-then" scenario (Jha and Gupta, 2003). Therefore considering these limitations, integrated optimal water resources allocation modelling should consider all the water users and equitably allocate water amongst users, whilst taking into consideration hydro-climatic variability. (Tsoukalas and Makropoulos

Published by Copernicus Publications on behalf of the International Association of Hydrological Sciences.

1972; Evers, 1992; Arranz and McCartney, 2007; Tilmant et al., 2010; Sharma et al., 2016). A variety of methodologies have been proposed and developed in order to embody in optimal water allocation the uncertainties of various influencing factors and hydro-system characteristics, such as available water flows, water demands, variations in water supplies, corresponding cost and benefit coefficients and policy regulations.

Most water resources systems decisions have to be made in the face of hydrologic uncertainty (Ajami et al., 2008). Rainfall, reservoir inflows, crop evapotranspiration are hydrologic variables which influence water resources decision making and all these variables are stochastic. Therefore application of optimization models for water management have to be formulated in a manner that considers the inherent hydrologic uncertainty in arriving at optimal water allocation decisions.

Stochastic Dynamic Programming (SDP) is a framework for modelling optimisation problems whose variables exhibit some uncertainty. It has been widely used by different researchers and professionals in optimization of reservoir operating policies with uncertainty in inflows.

Sharma et al. (2016) used the SDP approach to optimize the Ukai Reservoir in India. This SDP model which was based on the minimization of least squares sum of actual releases and storages from the respective targets was formulated by Buras (1985).

In the South-Eastern Lowveld of Zimbabwe, water is provided from five reservoirs namely Mutirikwi, Muzhwi, Bangala, Manjirenji and Siya on different river systems for sugarcane production. At present, all sugarcane for milling is grown under irrigation in the South-Eastern part of Zimbabwe. The current water allocation policy of using water permits has resulted in rising conflicts between the water authority and the users as it is a priority based policy. Also, due to the prolonged drought in the country, the reservoirs have been having very low volumes of water which has not been sufficient for irrigation of the crop throughout the past three years. In 2016, three reservoirs, Muzhwi, Bangala and Mutirikwi were temporarily decommissioned (reservoir has a minimum supply of domestic water for twenty one months) due to very low volumes and this put a lot of pressure on the remaining two reservoirs, the farmers and Runde Catchment Council (the authority responsible for water resource management in the basin) on how best to allocate the limited water resources. Thus an optimal multi-reservoir operating policy is needed whereby the releases from the five reservoirs meet existing water demands as much as possible, without emptying the reservoirs and without relying on one reservoir.

2 Study Area

The Lowveld of Zimbabwe falls under Runde catchment which is located in the South-Eastern part of Zimbabwe. Zimbabwe is divided into 7 catchment areas presented in Fig. 1, managed by seven Catchment Councils and 47 sub-Catchment Councils. The catchment is situated 21° S, 21° E and is 400 to 500 m above sea level. It has an area of approximately 41 000 km^2 which constitutes 22 % in area of Zimbabwe and 40 % of this catchment area is communal lands (Anderson et al., 1993). It lies in the dry parts of the country which cover Natural Regions III, IV and V with an average of 684 mm of rainfall per annum and temperatures as high as 43 °C are experienced (Meteorological Services Department (MSD)). The country is divided into five natural regions based on soil quality, rainfall regime, and vegetation amongst other factors. The land resource quality declines from Natural Region (NR) I through to NR V (Moyo, 2000; Vincent and Thomas, 1961). NR I receives average annual rainfall of more than 1000 mm, NR II between 700 and 1050 mm, NR III receives 500–800 mm, NR IV from 450–650 mm and NR V receives below 450 mm (Moyo, 2000; Vincent and Thomas, 1961). Approximately 90 % of the precipitation occurs during the period of mid-November to early April (MSD). The winter season is from April to August, and the hottest and driest period is from September to mid-November. The catchment is characterized by frequent droughts and an annual average evaporative demand of about 1500 mm (MSD). As a result of the combination of low rainfall and high evaporation, crop production is only possible with irrigation (Anderson et al., 1993).

Sugarcane production has been the mainstay of the Lowveld's economy. The sugar industry has produced significant foreign exchange for the national exchequer not to mention employment, ethanol, various industrial products, and the raw cane sugar which is consumed in large amounts in Zimbabwe. The sugar industry represents 1.4 % of national GDP, and a massive 95 % of Masvingo Province's accounted-for GDP (Scoones et al., 2017). Rainfall and runoff has an impact on the socio-economy of the people living in the Runde catchment and Zimbabwe as a whole hence the need to understand the temporal and spatial variability of both rainfall and runoff which determines the reliability of water supply.

3 Methodology

Observed hydrological and meteorological data; reservoir characteristics and operation rules; and water use for the catchment was used in the methodology. Figure 2 shows the flowchart of the methodology used in this study.

3.1 Assessment of precipitation and flow variability

The Mann-Kendall (MK) trend and the Wilcoxon signed rank tests were used to assess the presence of trends in the rainfall and inflow data from 1899 to 2015 on a monthly time scale. Data from the four rainfall stations indicated in Fig. 1 was used in the analysis. Eleven gauging stations were used

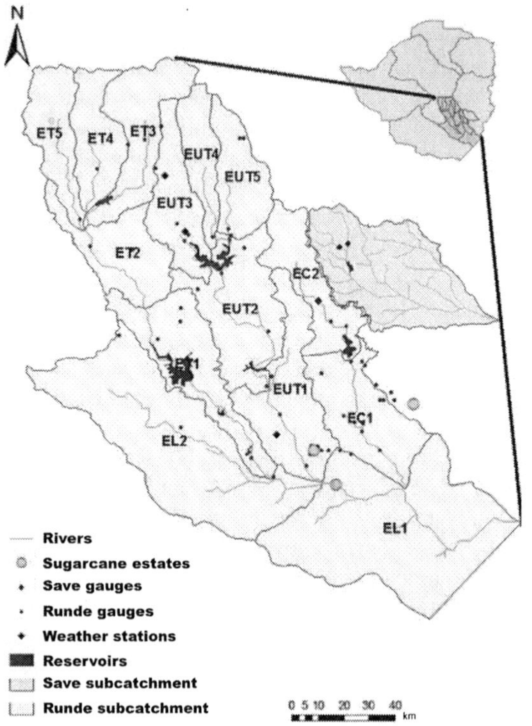

Figure 1. Study area map showing sugarcane estates and the water supplying reservoirs.

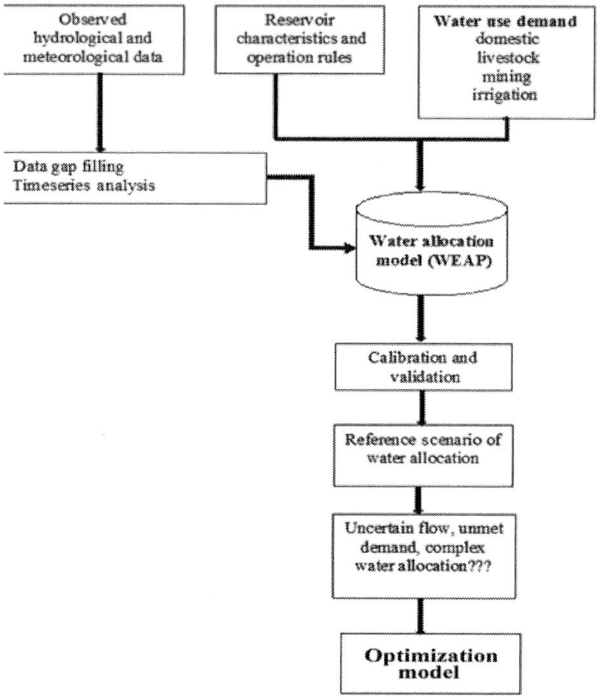

Figure 2. Methodology General Flow Chart.

as these are the ones which measure the inflow into the five reservoirs. Mann Kendall test was used as there is no need for the data to confirm to any particular distribution and it is less sensitive to sudden breaks due to time series which are inhomogeneous (Jaagus, 2006). To improve the robustness of the results the Wilcoxon Signed rank test was used in detection of significance of difference between two groups of data sets which were divided according to time periods so as compare the medians of two distributions.

3.2 Assessment of the current water allocation scenario for the catchment

The Water Evaluation and Planning System (WEAP) model was used to come up with a reference scenario for the 2015/2016 hydrologic year on the water allocation of the catchment and this was used as a baseline for optimization. A default scenario known as the "Reference Scenario" carries forward the Current Accounts data into the entire project period and serves as a point of comparison for other scenarios in which changes may be made to the system data. WEAP was used due to its basic principle of water balance accounting, priority based approach and user friendliness. The Zimbabwe National Water Act (1998) priorities were used in setting up the demand nodes. Water for primary needs is given the first priority, the environment comes second, Mining and Industries have a third priority, plantations such as fruit and forestry at the fourth, plantation-crops such as sugar cane are given the fifth priority and lastly seasonal crops have the sixth priority.

3.3 Optimisation Model (SDP Approach)

A backward recursive approach was used in coming up with the optimal policy. The SDP model developed was programmed in MATLAB® 2017. Different combinations of reservoir storage states (k) and inflows (i) were considered in coming up with the optimal policy for the reservoirs.

The optimal policies for each reservoir for each month were tabulated for different initial storage classes, inflows and end of month storage. The first-order Markov chain was used in computation of inflow transition probabilities (Vedula and Mujumdar, 2005) for the different months. The transition probabilities were used in the optimisation model represented by Fig. 3.

4 Results and Discussion

4.1 Variability Analysis for precipitation and flow

4.1.1 Trend test results for Rainfall

Precipitation trend tests showed a decreasing trend although not statistically significant. A negative Sen's slope from Mann Kendall test as well as more negative ranks compared

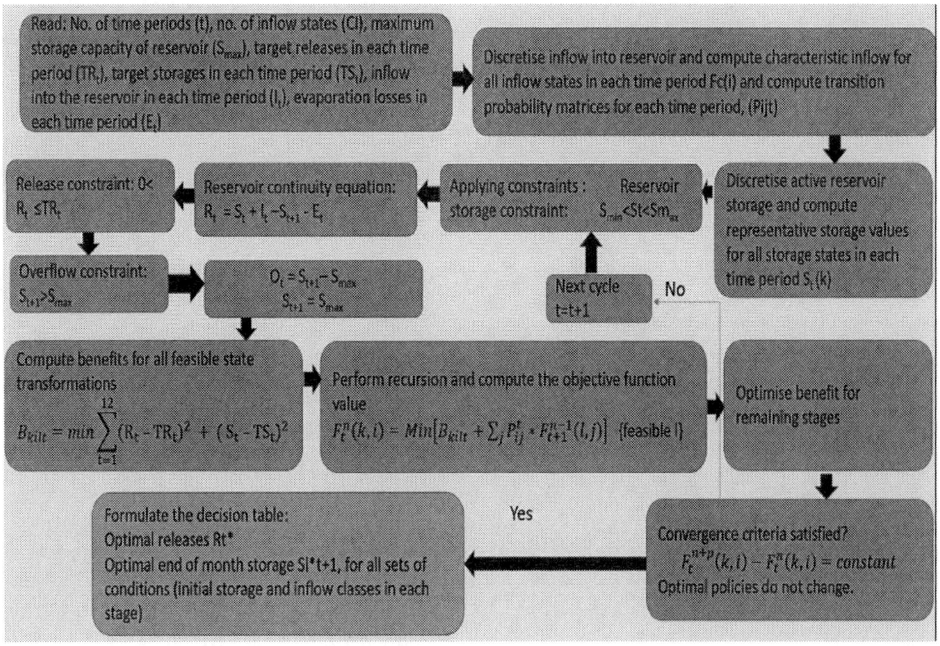

Figure 3. SDP Method Flow chart (Source: Sharma et al., 2016).

to positive ranks from Wilcoxon Signed Rank test were observed. This was predicted to cause a reduction in the runoff and hence the amount of available surface water due to a reduction in the reservoir volumes not only for the Lowveld, but across Zimbabwe.

4.1.2 Mann Kendall Trend and Wilcoxon Signed Rank Test Results for flows

There was need to understand the temporal and spatial variability of both rainfall and runoff which determines the reliability of water supply. Both tests showed that there were significant trends in the flows into all the five reservoirs from the eleven gauging stations. An overall decreasing trend was observed in the inflows, thereby a temporal variation experienced in the catchment over the years. The MK results showed nine out of eleven gauging stations having a negative Sen's slope. The Wilcoxon Signed Rank Test results also showed nine of the eleven stations having more negative ranks than positive ranks. Thus there was a significantly decreasing trend in the flows.

4.2 Assessment of the current water allocation scenario for the catchment

Table 1 shows the different total annual demand and total unmet annual demand of the subcatchment irrigation demand sites for each individual subzone in the study area map in Fig. 1. A first order validation of the WEAP model was done comparing the simulated flows to the observed flows. The

Table 1. 2015/2016 hydrologic year total demand and unmet demand for Irrigation.

Irrigation Demand site	Total Demand (cumecs)	Total Unmet Demand (cumecs)
EC1	112.80	91.93
EC2	292.49	238.38
EL1	282.00	157.99
EL2	52.92	11.64
ET1	12.16	6.35
ET2	40.64	21.25
ET3	143.21	93.90
ET4	3.53	1.85
ET5	0.90	0.47
EUT1	240	134.48
EUT2	1550.08	1199.64
EUT3	2.8	2.13
EUT4	1.2	0.91
EUT5	7.42	5.65
Total	2742.13	1966.579

Nash-Sutcliffe efficiency (NSE) was used as an indicator of the model performance. The NSE for the different river systems ranged from 0.66–0.99. A large monthly variation in irrigation water demands existed with the dry season of July to November having the highest unmet demand. Unmet demand is the amount of water not satisfied to the user (shortage).

The high unmet demands could be due to the decommissioning of the three major reservoirs which supply irrigation

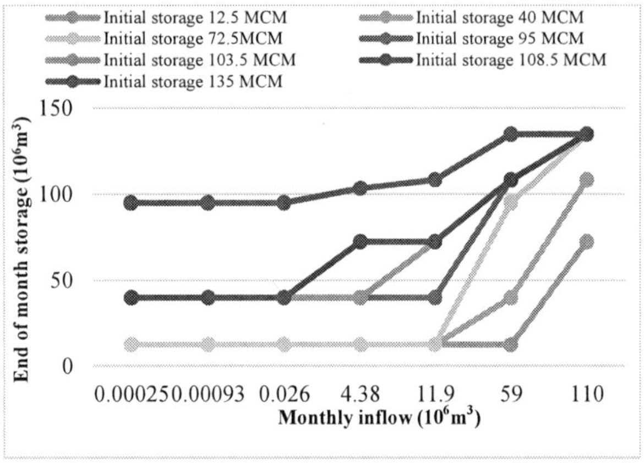

Figure 4. Muzhwi March Rule curve.

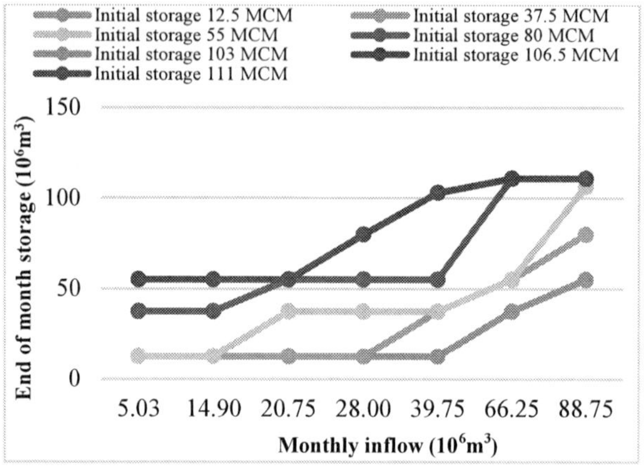

Figure 5. Siya February Rule Curve.

water to the estates and also due to the current water allocation priorities of the water authority.

4.3 (SDP) Optimal Policy

Figure 4 shows that in March, the target release was 32.94 MCM whilst the target storage was expected to be the full reservoir capacity of 158 MCM because this will be towards the end of the rainy season to ensure maximum efficiency of the project in meeting conservation demands during the lean period.

For Siya reservoir, in February, the target release was 38.92 MCM and storage was 90 MCM. The target storage is not met for most of the different combinations. The demands are mostly met by the different combinations of initial storages and the inflows as shown in Fig. 5.

From the five different reservoirs, for all the different combinations of initial storage states and inflows coming into the reservoirs, the model tries its best to get a balance between the target storage and the target release. All the results from the different reservoirs show that at any particular moment, there is a possible amount of water that can be released in trying to meet the target release and at the same time the reservoir would not be emptied. All the minimum storages were above the decommissioning level (twenty-one months' supply of primary water use), thus it can be used as a flexible tool for allocation tool in the semi-arid catchment which has competing and conflicting uses.

Operating one reservoir at a time might not fully satisfy all the desired targets but if water can be released from all the different reservoirs depending on the initial state of storage and the inflow received, the demands can be fully met.

5 Conclusions

There was no statistically significant trend in monthly rainfall for the period 1899–2010 but there was a statistically significant decreasing trend for the monthly reservoir inflows from 1959–2015.

The current water allocation method has huge deficits within the catchment which compromises the crop yield and thereby economic benefits not recognised as expected. The irrigation unmet demand (approximately 70 % of the total annual demand) is the largest due to the priority based allocation approach and therefore there is need for a better water allocation approach in the study area which can optimise benefits to the different water users.

The operation policies derived from the SDP model for each month and for each reservoir are more stable and adaptable for release decisions due to fluctuations in the inflows into the reservoir.

A crop yield optimisation approach for the study area may yield better results as it addresses the optimal yield that can be obtained depending on the amount of water released at any particular time. This may consider the critical stages of the crop growth which have to be given priority so as to maximise the crop yield.

The proposed rule curves can serve as a decision-making tools for long-term operation of the reservoirs, thereby maximizing the benefits derived from the reservoir.

Data availability. The data used in the research can be accessed on request from the researchers.

Competing interests. The authors declare that they have no conflict of interest.

Special issue statement. This article is part of the special issue "Understanding spatio-temporal variability of water resources and the implications for IWRM in semi-arid eastern and southern Africa". It is a result of the IAHS Scientific Assembly 2017, Port Elizabeth, South Africa, 10–14 July 2017.

Acknowledgements. The first author wishes to acknowledge the financial support received from WaterNet Organisation.

Edited by: Hodson Makurira
Reviewed by: Evison Kapangaziwiri and one anonymous referee

References

Ajami, N. K., Hornberger, G. M., and Sunding, D. L.: Sustainable water resource management under hydrological uncertainty, Water Resour. Res., 44, 1–10, 2008.

Anderson, I. P., Moyo, M., Nyamwanza, B., and Brinn, P. J.: Physical resource inventory of the communal lands of Zimbabwe-An Overview, NRI Bulletin 60, Chatham, UK: Natural Resource Institute, 1993.

Arranz, R. and Mccartney, M.: Application of the Water Evaluation And Planning (WEAP) Model to Assess Future Water Demands and Resources in the Olifants Catchment, South Africa (No. H040224), International Water Management Institute, available at: http://www.iwmi.cgiar.org/Publications/Working_Papers/working/WOR116.pdf, (last access: 31 January 2018), 2007.

Buras, N.: An application of mathematical programming in planning surface water storage, J. Am. Water Resour. As., 21, 1013–1020, 1985.

Evers, A. J. M.: Optimizing intra- and inter-seasonal water allocation from an irrigation reservoir subject to stochastic inflows, A Thesis Submitted for the Degree of Doctor of Philosophy, Oklahoma State University, 1992.

Jaagus, J.: Climatic Changes in Estonia during the Second Half of the 20th Century in Relationship with Changes in Large-Scale Atmospheric Circulation, Theor. Appl. Climatol., 83, 77–88, 2006.

Jha, M. and Gupta, A. D.: Application of Mike Basin for Water Management Strategies in a Watershed, Water Int., 28, 27–35, 2003.

Juízo, D. and Lidén, R.: Modeling for transboundary water resources planning and allocation: the case of Southern Africa, Hydrol. Earth Syst. Sci., 14, 2343–2354, https://doi.org/10.5194/hess-14-2343-2010, 2010.

Moyo, S. (Ed.): Zimbabwe environmental dilemma: balancing resource inequities, Harare, Zimbabwe Environmental Research Organisation, 161 pp., 2000.

Mujumdar, P. P. and Nirmala, B.: A Bayesian stochastic optimization model for a multi-reservoir hydropower system, Water Resour. Manag., 21, 1465–1485, https://doi.org/10.1007/s11269-006-9094-3, 2007.

Namchaiswadwong, K., Bhaktikul, K., and Kongjun, T.: Optimization Of Multiple Reservoir Releases Using Genetic Algorithms: Case Study Of Mae Klong River Basin, Thailand, 5th Thaicid National Symposium, p. 10, Beijing, THAICID, 2004.

Reddy, M. J. and Kumar, D. N.: Optimal reservoir operation for irrigation of multiple crops using elitist-mutated particle swarm optimization, Hydrolog. Sci. J., 52, 686–701, https://doi.org/10.1623/hysj.52.4.686, 2007.

Scoones, I., Mavedzenge, B., and Murimbarimba, F.: Sugar, People and Politics in Zimbabwe's Lowveld, J. S. Afr. Stud., 43, 567–584, https://doi.org/10.1080/03057070.2016.1187972, 2017.

Sert, M., Opan, M., and Temiz, T.: Multiobjective Optimal Planning in Multiple Reservoir Systems, International Congress on River Basin Management, 554–567, 2007.

Sharma, P. J., Patel, P. L., and Jothiprakash, V.: Efficient discretization of state variables in stochastic dynamic programming model of Ukai reservoir, India, ISH J. Hydraul. Eng., 22, 293–304, https://doi.org/10.1080/09715010.2016.1204632, 2016.

Tilmant, A., Kinzelbach, W., Beevers, L., and Juizo, D.: Optimal Water Allocation in the Zambezi Basin, 9th International Congress on Environmental Modelling and Software, 188, available at: https://scholarsarchive.byu.edu/iemssconference/2010/all/188 (last access: 31 January 2018), 2010.

Tsoukalas, I. and Makropoulos, C.: Hydrosystem Optimization With the Use of Evolutionary Algorithms?: the Case of Nestos River, 13th International Conference on Environmental Science and Technology (CEST2013), At Athens, Greece, 1972.

Vedula, S. and Mujumdar, P. P.: Water Resources Systems, New Delhi, India, Tata McGraw-Hill Education, 2005.

Vincent, V. and Thomas, R. G.: An agro-ecological survey of Southern Rhodesia: Part I agro-ecological survey, Salisbury, Government Printers, 1961.

Zimbabwe Water Act: chap. 20:24, Harare, Government Printers, available at: www.parlzim.gov.zw/acts-list/water-act-20-24, (last access: 31 January 2018), 1998.

https://doi.org/10.5194/piahs-378-73-2018

Water management challenges at Mushandike irrigation scheme in Runde catchment, Zimbabwe

Jose A. Malanco[1,2], **Hodson Makurira**[1], **Evans Kaseke**[3], **and Webster Gumindoga**[1]

[1]Department of Civil Engineering, University of Zimbabwe, P.O. Box MP167, Mt. Pleasant, Harare, Zimbabwe
[2]National Directorate of Water, Ministry of Public Works and Housing, P.O. Box 1613, Rua da Argelia 163, Maputo, Mozambique
[3]Zambezi River Basin Commission, 128 Samora Machel Avenue, Harare, Zimbabwe

Correspondence: Webster Gumindoga (wgumindoga@gmail.com)

Received: 23 November 2017 – Revised: 11 February 2018 – Accepted: 14 February 2018 – Published: 29 May 2018

Abstract. Mushandike Irrigation Scheme, constructed in 1939, is located in Masvingo District and is one of the oldest irrigation schemes in Zimbabwe. Since 2002, the scheme has experienced severe water shortages resulting in poor crop yields. The low crop yields have led to loss of income to the smallholder farmers who constitute the irrigation scheme leading to water conflicts. The water stress at the scheme has been largely attributed to climate change and the uncontrolled expansion of the land under irrigation which is currently about 1000 ha against a design area of 613 ha. This study sought to determine the actual causes of water shortage at Mushandike Irrigation Scheme. Hydro-climatic data was analysed to establish if the Mushandike River system generates enough water to guarantee the calculated annual yield of the dam. Irrigation demands and efficiencies were compared against water availability and dam releases to establish if there is any deficit. The Spearman's Rank Correlation results of 0.196 for rainfall and 0.48 for evaporation confirmed positive but insignificant long-term changes in hydro-climatic conditions in the catchment. Water budgets established that the yield of the dam of 9.2×10^6 m^3 $year^{-1}$ is sufficient to support the expanded area of 1000 ha provided in-field water management efficiencies are adopted. The study concludes that water shortages currently experienced at the scheme are a result of inefficient water management (e.g. over-abstraction from the dam beyond the firm yield, adoption of inefficient irrigation methods and high channel losses in the canal system) and are not related to hydro-climatic conditions. The study also sees no value in considering inter-basin water transfer to cushion the losses being experienced at the scheme.

1 Introduction

In most regions of the world, over 70 % of freshwater is used for agriculture (World Bank, 2017). This figure can be as high as 85 % in low and middle income countries where agriculture is the major economic activity (World Bank, 2006). By 2050, feeding a planet of 9 billion people will require an estimated 50 % increase in agricultural production and a 15 % increase in water withdrawals (World Bank, 2017).

In Africa, where agriculture is the backbone of most of the continent's economies, providing about 60 % of all employment, huge investments in infrastructure for irrigation schemes over the last half century have been made, highlighting the importance of agriculture in food security for the growing population (Tafesse, 2003). Although the achievements of irrigation in ensuring food security and improving rural welfare have been impressive, experience also indicates problems and failures of irrigated agriculture (FAO, 2001; Cai et al., 2003). The poor performance of irrigation schemes are characterized by inefficient water use, unreliable water supply, degradation of irrigation infrastructure, lack of sound technical extension facilities and unconvincing farm management practices (Moyo et al., 2017; Malano and Burton, 2001). These will be compounded by uncertainties related to climate change (Nam et al., 2017).

Published by Copernicus Publications on behalf of the International Association of Hydrological Sciences.

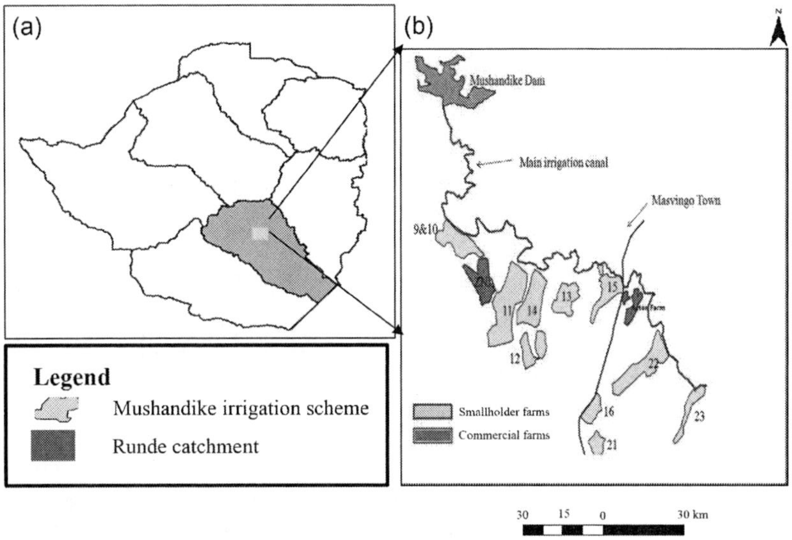

Figure 1. **(a)** Location of Mushandike Irrigation Scheme **(b)** Layout of Mushandike Irrigation Scheme.

In Zimbabwe, the total irrigated land dropped from 173 513 ha to approximately 120 000 ha primarily as a consequence of the land reform programme (Nhundu and Mushunje, 2010) and general economic challenges in the country. Efforts are being made to improve agricultural productivity in the communal lands through development and resuscitation of irrigation schemes where farmers can grow crops during both the wet and dry seasons (Makombe et al., 1998).

The Mushandike Irrigation Scheme, located 20 km south of Masvingo Town in Zimbabwe, is one of Zimbabwe's oldest communal irrigation schemes which provides livelihood income to the participating farmers and, also, is a source of fresh vegetables for the historical town of Masvingo. However, water shortages, blamed on low dam levels in the last 20 years have led to poor performance of the irrigation scheme resulting in food insecurity in the region and loss of livelihood income for the irrigation scheme participants (The Herald, 2017).

This study sought to confirm if the changes in the hydrological regime of the Mushandike Dam catchment have led to the serious water shortages that are attributed to the poor performance of the irrigation scheme.

2 Material and methods

2.1 Study Area

The study area is the Mushandike Irrigation Scheme, situated in the Tokwe sub-catchment within the Runde Catchment of Zimbabwe (Fig. 1). The climate is characterized by an unreliable rainfall season between November to April with the winter cropping season being experienced between May and September. About 90 % of the rainfall occurs during the period mid-November to early April. The average rainfall is 520 mm $year^{-1}$ with frequent and often extended dry spells during the season. The mean annual temperature is about 21 °C.

The catchment area upstream of Mushandike Dam is 319 km^2 and the irrigation scheme lies immediately downstream of the dam. The scheme was established in 1939 to irrigate 613 ha (AGRITEX, 1989). The system flood irrigation methods with water being distributed through unlined canals. However, membership has increased and the area under irrigation has now expanded to 997 ha. At design stage, assumption of 1.5×10^3 m^3/ha/year was made as the annual irrigation demand. This is the general guideline used for the design of small-scale irrigation systems in Zimbabwe. With improvements in farming systems over time, this allocation is considered to promote inefficiency and, hence, there is room for increasing the area under irrigation if more efficient irrigation methods are adopted.

2.2 Research approach

2.2.1 Analysis of rainfall and evaporation

Annual rainfall, evaporation and temperature data was analysed using data recorded at two stations in Masvingo Town and at Mutirikwi Dam.

2.2.2 Analysis of storage levels in the dam

There is no streamflow measuring station upstream of the Mushandike Dam. However, there is data available on storage levels and releases from the dam. From this, a water balance at the dam could be conducted.

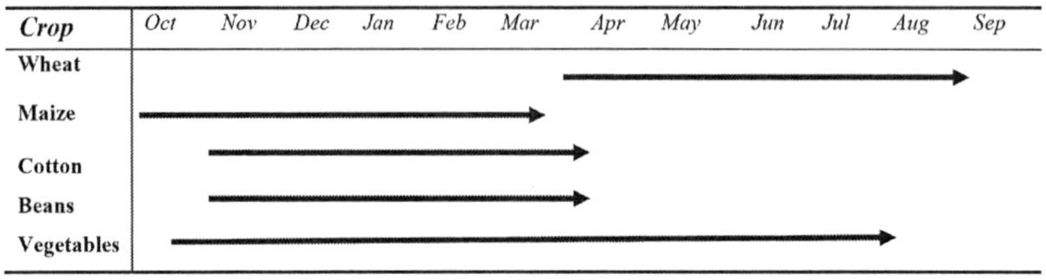

Figure 2. Typical cropping pattern and calendar at Mushandike Irrigation Scheme.

2.2.3 Hydro-climatic trend analysis

Monthly rainfall data from the Masvingo rainfall station between 1989 and 2016 was collected. The Spearman's rank correlation method was used to analyse the trends in the time series of hydro-climatic data collected. This analysis would show if there are any visible trends in rainfall amounts that may be causing reduced inflows into the dam. Similar analyses were conducted for temperature since crop water requirements are strongly related to prevailing temperature. The Spearman's correlation coefficient is calculated from Eq. (1):

$$R_{\text{sp}} = 1 - \frac{6\sum D_i^2}{n(n^2-1)} \qquad (1)$$

Where, R_{sp} is the Spearman's Rank Correlation coefficient, n is the length of data series, and D_i is the difference between the rankings. A positive R_{sp} value indicates increasing trend while a negative value indicates decreasing trends in annual rainfall amounts received. A value of zero indicates no change in trends.

The precipitation variation from the mean was calculated using the Standardized Precipitation Index (S_{pi}) calculated using Eq. (2):

$$S_{\text{pi}} = \frac{P - P_m}{\delta} \qquad (2)$$

Where, S_{pi} is the standardized precipitation index, P is the precipitation (mm year^{-1}) in a given year; P_m is average annual precipitation (mm year^{-1}) and δ is the standard deviation.

The crop water requirements were estimated from the FAO pan evaporation method (FAO, 1992) as given by Eq. (3):

$$\text{ET}_{\text{c}} = K_{\text{c}} \cdot \text{ET}_o \qquad (3)$$

where, ET_{c} is crop water requirement (mm day^{-1}); K_{c} is crop coefficient (dimensionless); and ET_o is the reference crop evapotranspiration (mm day^{-1}). The method was applied for maize (200 ha), beans (200 ha) and cotton (200 ha) which are grown as summer crops, wheat (600 ha) which is grown as a winter crop and vegetables (300 ha) which are grown throughout the year (Fig. 2).

Crop coefficients for each crop and for different growth stages were adopted from the FAO guidelines. Water requirements were calculated for each crop on a monthly basis.

3 Results and discussion

3.1 Analysis of rainfall and evaporation

Both the rainfall and evaporation patterns show an increasing visual trend (Fig. 2). Similar observations were made for temperature. The combination of both trends would naturally be linked to increased water demands for crops.

3.2 Analysis of inflows and storage

Since 1978, the storage levels do not show any period when the dam reached full supply capacity (FSC) (Fig. 5). However, the abstractions from the dam have been exceeding the design yield of 9.2×10^6 m^3 year^{-1} since 2006. This is not recommended as it affects the reliability of water security in the long term.

3.3 Hydro-climatic trends and analysis

The Spearman's rank-correlation method, revealed weak positive rainfall and evaporation trends of 0.19 and 0.48), respectively. Both trends are insignificant, i.e. have $p \geq 0.05$. However, pronounced cyclic changes particularly for rainfall were observed which are characterised by longer dry periods and shorter wet periods (Fig. 6).

The annual irrigation water requirement calculated using the FAO crop water requirement estimation method for the expanded scheme of 997 ha is 8×10^6 m^3 year^{-1} assuming the same cropping pattern is maintained over the years. The dam yield is 9.2×10^6 m^3 year^{-1}. The available water, based on the calculated firm yield, is, therefore, sufficient water to irrigate the expanded irrigation area.

4 Discussion of findings

The Mushandike Irrigation Scheme in Zimbabwe is experiencing serious water stress which has been attributed to increased area under irrigation and reduced water availability due to climate change. These are similar reasons advanced to explain underperformance of many small-scale irrigation schemes across the country. However, the study has established that the hydroclimatic conditions in the Mushandike Dam catchment have not changed significantly. The water

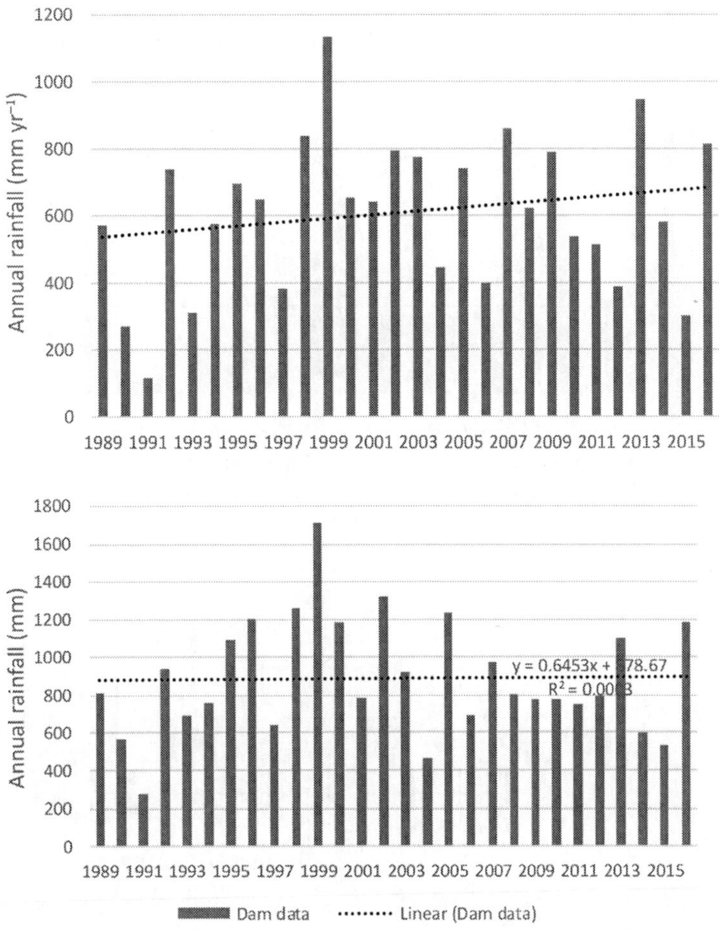

Figure 3. Rainfall trends at Mushandike Irrigation Scheme.

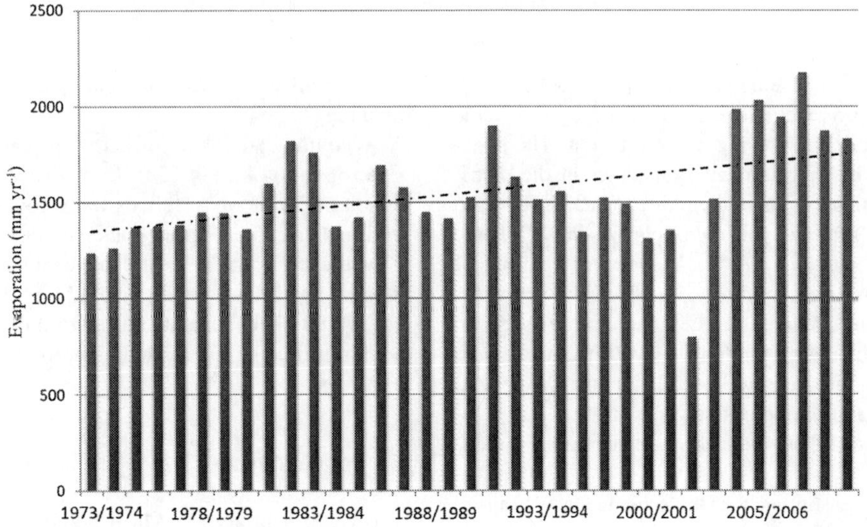

Figure 4. Evaporation trends at Mushandike Irrigation Scheme.

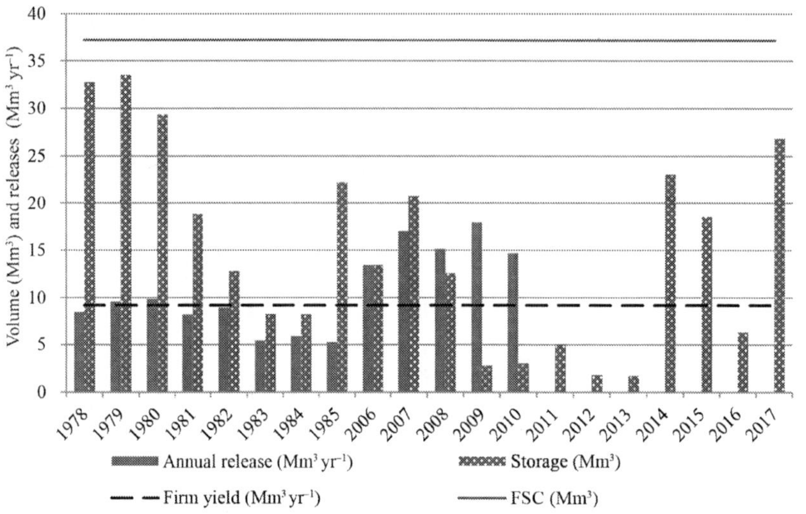

Figure 5. Storage levels in Mushandike Dam.

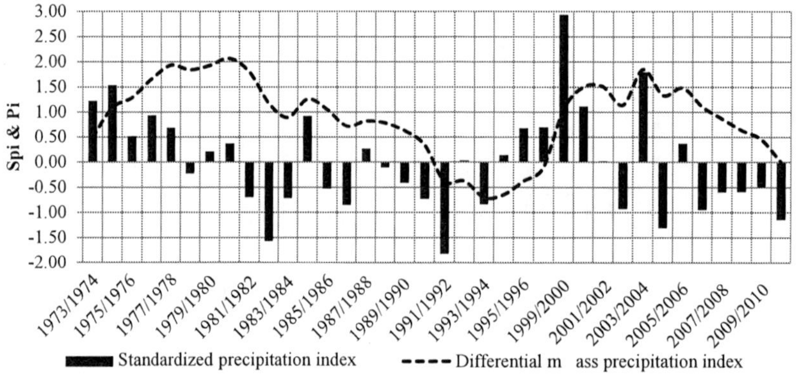

Figure 6. Cyclic changes on observed Rainfall (1974–2011).

stress at the irrigation scheme can only be attributed to in-field water application inefficiencies in the irrigation scheme itself. Flood irrigation methods are not efficient. The situation is worsened by high transmission losses in the distribution system which comprise of unlined canals. The crop water requirements for the typical crops grown in the scheme have been estimated to be 8×10^3 m^3/ha/year yet the desgn of the scheme assumed allocations of 15×10^3 m^3/ha/year. This has nurtured a culture of inefficiencies amongst the users and operators. Irrigation design guidelines for Zimbabwe should be adjusted to acknowledge improved irrigation and general farming methods and the increasing value of water that is associated with increased competition and uncertainties in climate conditions.

The serious water shortages at the scheme started during the period when annual abstractions started to exceed the design yield of the dam. This also suggests that the operation rules for the dam were violated in recent years. This, obviously, leads to increased insecurity of supply from the dam on a long-term basis.

Accommodating the 997 ha of irrigation land using the generous allowance of 15×10^3 m^3/ha/year would generate a deficit of 7 Mm^3 year^{-1}. This water could be divered from a different catchment through inter-basin transfer. Technically, it would be possible as the adjacent Tokwe River system records good flows on an annual basis. However, this strategy would likely generate upstream-downstream conflicts in addition to being interpreted as promoting inefficient water use.

5 Conclusions

The water shortages at Mushandike Irrigation Scheme are caused by infield inefficiencies and not by the hydrology of the upstream catchment. The original irrigation design assumed a water allocation of 15×10^3 m^3/ha/year which, in itself, is more than generous and promotes inefficient use of scare water resources. Inter-basin water transfer, while it is an option, cannot be promoted now as such intervention would be interpreted as promoting inefficient water use. Operation rules for the dam, which are related to the inflow patterns and the design yield of the dam should be respected at all times. Over-abstraction from the dam leads to long term water shortage. The problem of water stress at the scheme can be solved by improving water allocations and minimising in-field losses without necessarily considering inter-basin water transfers.

Data availability. Data are not publicly accessible due to restrictions imposed by the supplier of the data.

Competing interests. The authors declare that they have no conflict of interest.

Special issue statement. This article is part of the special issue "Understanding spatio-temporal variability of water resources and the implications for IWRM in semi-arid eastern and southern Africa". It is a result of the IAHS Scientific Assembly 2017, Port Elizabeth, South Africa, 10–14 July 2017.

Acknowledgements. The authors greatly appreciate WaterNet Phase 3 for funding this research.

Edited by: Christophe Cudennec
Reviewed by: Hubert H. G. Savenije and one anonymous referee

References

AGRITEX: Mushandike resettlement scheme project report update, Ministry of Lands, agriculture and rural resettlement, Harare, Zimbabwe, 1989.

Cai, X., McKinney, D. C., and Rosegrant, M. W.: Sustainability analysis for irrigation water management in the Aral Sea region, Agr. Syst., 76, 1043–1066, 2003.

FAO: Irrigation Water Management, Training Manual No. 7, FAO, Land and Water Development Division, Rome, Italy, 1992.

FAO: Smallholder irrigation technology: Prospects for Sub-saharian Africa, International Programme for Technology and Research in Irrigation and Drainage Knowledge Synthesis Report No. 3, March 2001, Melvyn Kay FAO/IPTRID Consultant IPTRID Secretariat Food and Agriculture Organization of the United Nations Rome, 2001.

The Herald: available at: http://www.herald.co.zw/mushandike-irrigation-scheme-under-threat/, last access: 26 October 2017.

Makombe, G., Makadho, J. M., and Sampath, R. K.: An analysis of the water management performance of small holder irrigation schemes in Zimbabwe, Irrigation and Drainage Systems, 22, 253–263, 1998.

Malano, H. and Burton, M.: Guidelines for benchmarking performance in the irrigation and drainage sector, Food and Agriculture Organization of the United Nations, Rome, 2001.

Moyo, M., van Rooyen, A., Moyo, M., Chivenge, P., and Bjornlund, H.: Irrigation development in Zimbabwe: Understanding productivity barriers and opportunities at Mkoba and Silalatshani irrigation schemes, Int. J. Water Resour. D., 33, 740–754, https://doi.org/10.1080/07900627.2016.1175339, 2017.

Nam, W., Kim, T., Hong, E., and Choi, J.: Regional Climate Change Impacts on Irrigation Vulnerable Season Shifts in Agricultural Water Availability for South Korea, Water, 9, 735, https://doi.org/10.3390/w9100735, 2017.

Nhundu, K. and Mushunje, A.: Analysis of irrigation development post fast track land reform programme, A case study of Goromonzi District, Mashonaland East Province, Zimbabwe, A Poster presented at the Joint 3rd African Association of Agricultural Economists (AAAE) and 48th Agricultural Economists Association of South Africa (AEASA) Conference, Cape Town, South Africa, 19–23 September 2010.

Tafesse, M.: Small scale irrigation for food security in sub- Saharan Africa, The ACP-EU Technical Centre for Agricultural and Rural Cooperation (CTA), Ethiopia, 2003.

World Bank: Reengaging in Agricultural Water Management, Challenges and Options, The World Bank Washington, DC, 2006.

World Bank: available at: https://blogs.worldbank.org/opendata/chart-globally-70-freshwater-used-agriculture (last access: 3 March 2018), 2017.

https://doi.org/10.5194/piahs-378-79-2018

Open Access
PIAHS

Water use dynamics of young and mature apple trees planted in South African orchards: a case study of the Golden Delicious and Cripps' Pink cultivars

Zanele Ntshidi[1,2], **Sebinasi Dzikiti**[2], **and Dominic Mazvimavi**[1]

[1]Department of Earth Sciences, University of the Western Cape, Bellville 7535, South Africa
[2]Council for Scientific and Industrial Research, Natural Resources and Environment, Stellenbosch 7600, South Africa

Correspondence: Zanele Ntshidi (zntshidi@gmail.com)

Received: 10 November 2017 – Accepted: 22 December 2017 – Published: 29 May 2018

Abstract. Apple orchards have previously been bearing 60–80 $t\,ha^{-1}$ at most. However in recent years yield has increased to more than 100 $t\,ha^{-1}$. There is need to understand the water requirements of the high yielding orchards, given that high crop loads are associated with high water use rates. The aim of this study was to quantify the water requirements of young and mature unstressed apple orchards. We also assess the impact of climate variables on transpiration rates. Data was collected in 4 orchards in the Western Cape Province. The orchards comprised young non-bearing (< 3 years) and mature trees planted to the Golden Delicious and Cripps' Pink cultivars, all under micro-sprinkler irrigation. Transpiration by the trees was measured using heat pulse velocity sap flow sensors hourly throughout the growing season (October–June). Weather was monitored using an automatic weather station. Tree transpiration was linearly related to the solar radiation, but the relationship with the vapour pressure deficit (VPD) was non-linear. There were no significant differences ($p>0.05$) in the sapflux density of the Golden Delicious and Cripp's Pink cultivars. This suggests that these two cultivars have similar water use characteristics. Mature orchards transpired between 6000 to 8000 $m^3\,ha^{-1}\,season^{-1}$ while non-bearing orchards used between 2000 to 3000 $m^3\,ha^{-1}\,season^{-1}$.

1 Introduction

South Africa is currently the 7th biggest exporter of apples (*Malus domestica* Borkh.) in the world. The main Southern hemisphere competitor is Chile, which is ranked third in terms of export. In South Africa, most apples are grown in the Eastern and Western Cape Provinces. The Western Cape experiences Mediterranean climatic conditions and most of the rain falls outside the fruit growing season in winter. This region is projected to experience severe water shortages in future due to the rapidly increasing population, increasing agricultural and industrial activities and climate change (Midgley and Lötze, 2011). Accurate information on water use by apple orchards is therefore critical for the sustainability of the multi-billion Rand deciduous fruit industry. This information will be useful for, among other things: (1) developing irrigation scheduling guidelines for the apple industry, (2) water allocation and planning, and; (3) developing strategies for coping with drought whose frequency and severity is expected to increase in the near future in the South Western parts of the country. Severe water deficits imposed by a lack of irrigation not only reduce the current year's yields, but they also negatively affect production in subsequent years, enhance alternate bearing and damage or kill trees (Fereres et al., 2012). An important information gap currently exists regarding the water requirements of high yielding ($\geq 100\,t\,ha^{-2}$) apple orchards which have become the norm in this region in recent years. There is also no published information on how water use varies with orchard growth stages from planting until full-bearing. Although several studies have quantified water use by apple orchards (Braun et al., 2000; Volschenk et al., 2003; Dragoni et al., 2005; Gush and Taylor, 2014) these studies were not done in exceptionally high yielding orchards. The specific objective of the study was to quan-

Published by Copernicus Publications on behalf of the International Association of Hydrological Sciences.

tify how transpiration of apple orchards change with canopy cover, to facilitate the development of irrigation guidelines. An example of a programme in which the fruit industry could directly benefit from the data generated in this study is the recently launched Validation and Verification of Lawful Water Use (V&V) programme. This programme has been rolled out throughout South Africa by the Department of Water and Sanitation (DWS) mainly to confirm how much water may be used lawfully for commercial purposes (including irrigation) to facilitate water use licensing in line with the country's Water Act of 1998. Therefore, this study sets a benchmark on the water requirements of the country's most productive apple orchards.

2 Materials and Methods

2.1 Study sites

Four orchards (two non-bearing and two full-bearing) were monitored in the Koue Bokkeveld (KBV) region during the 2014/15 growing season. Yield in all full-bearing orchards was consistently above 100 t ha^{-1}. The full-bearing "Cripps' Pink" and "Golden Delicious" orchards were in adjacent blocks KP06 and KP07 at Kromfontein Farm in the Koue Bokkeveld (S32.97120°; E019.23295°; 876 m a.s.l.). They were both ca. 6 ha in size and aged 22 years for the "Golden Delicious" and 9 years for the "Cripps' Pink". They had similar soil types namely the deep sandy Fernwood soils with approximately 30 % stone. Tree spacing in both orchards was 4 m × 1.5 m giving a plant density of 1667 trees per hectare. Rootstock in both orchards was the industry standard M793. Both orchards were irrigated with medium range micro sprinkler systems.

The non-bearing "Golden Delicious" orchard was at Lindeshof Farm (S32.95002°; E019.20737°; 820 m a.s.l.), less than 2 km from Kromfontein. Both farms belong to Dutoit Agri and therefore they had similar management practices. The non-bearing "Golden Delicious" orchard was two years old and it was about 3.17 ha in size. The rootstock was also the M793 and the trees were planted in rows with a north-south orientation. Tree spacing and soil types were similar to that of the full bearing orchards. The non-bearing "Cripps' Pink" orchard was at Paardekloof farm (S32°15′39.02″; E019°16′00.74″; 900 m a.s.l.) on MM109 rootstock. The Paardekloof orchard was about 6.3 ha, but trees were planted at 3.5 m × 1.25 m giving a substantially higher plant density of 2285 trees per hectare. The trees were three years old and also growing on deep sandy soils irrigated with a medium range sprinkler system.

2.2 Climate data

Weather data was measured using an automatic weather station in close proximity to the orchards. The station was installed at a site with uniform short grass cover whose attributes were close of those of the grass reference surface (Allen et al., 1998). The equipment comprised a pyranometer (Model: SP 212 Apogee Instruments, Inc., Logan UT, USA), which measured the solar irradiance and it was installed on a north facing horizontal levelling fixture to avoid self-shading and to minimize cosine errors. Air temperature and relative humidity were measured using a temperature and humidity probe (Model: HMP60 Campbell Scientific, Inc., Logan UT, USA) installed at a height of about 2.0 m above the ground. Rainfall was recorded using a tipping bucket rain gauge (Model: TE525-L; Campbell Scientific, Inc., Logan UT, USA). All the sensors were connected to a data logger (Model: CR1000 Campbell Scientific, Inc., Logan UT, USA) programmed with a scan interval of 10 s and the output signals were processed at hourly and daily intervals, respectively.

2.3 Tree water use

Transpiration in each of the full-bearing orchards was measured from September 2014 until June 2015 on six trees of different stem sizes using the heat ratio method of the heat pulse velocity (HPV) sap flow technique (Burgess et al., 2001). Four sets of heater probes and T-type thermocouple pairs were inserted into the sap wood of the stems at depths ranging from 10 to 50 mm under the bark, depending on the size of the stem, to account for the circumferential and radial variation in sap velocity (Wullschleger and King, 2000). The HPV data was corrected for wounding, moisture fraction and wood density at the end of the experiment according to the approach by Swanson and Whitfield (1981). The wound widths were measured at five positions across the length of the wound created by sensor implantation using Vernier callipers. The conducting sapwood area was determined by injecting a weak solution of methylene blue dye into the stems towards the end of the experiment to determine the extent of the active xylem vessels. The bark-sapwood-heartwood boundaries were clearly distinguishable with this method. Transpiration by the smaller trees in the non-bearing orchards was measured using Granier probes (Model: TDP 10, Dynamax Inc., Houston, USA) (Granier, 1987). Three trees were instrumented per orchard and the average sap velocity was determined in the depth range 0 to 10 mm of the stems. The sensors were installed at a height between 50 and 75 cm from the ground to eliminate errors due to cold sap especially early in the morning. Installing the sensors at the recommended 1.0 m height above the ground was not feasible because the trees branched close to the ground. A double layer of aluminium bubble wrap was wrapped around the sensors to minimize the effects of exogenous heating on the sap temperature signals. The TDP sensors were calibrated against the more accurate stem heat balance sap flow sensors given the tendency of TDPs to underestimate transpiration rates (Steppe et al., 2010). Both the TDP and HPV sap flow

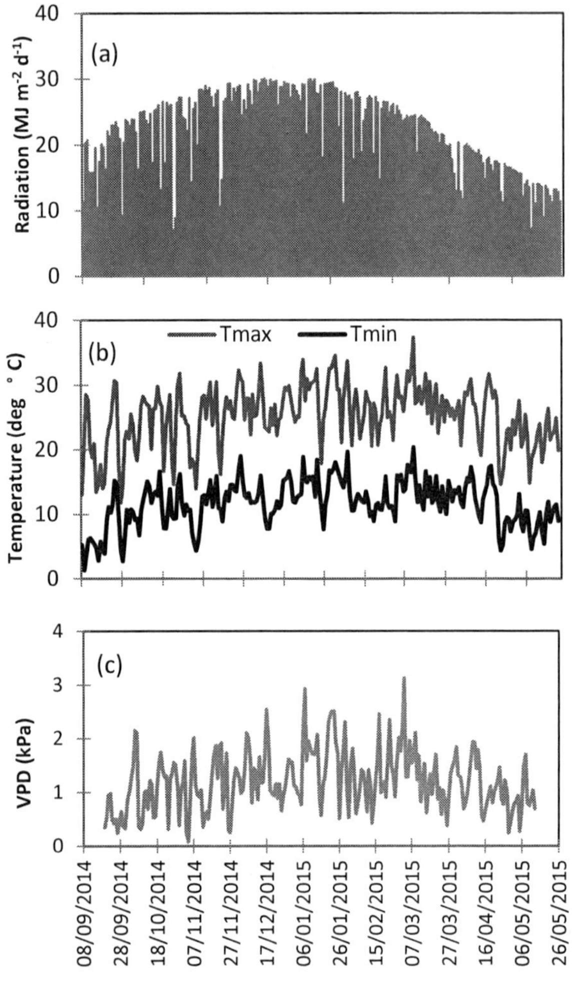

Figure 1. **(a)** Solar radiation, **(b)** minimum and maximum temperature and **(c)** Vapour Pressure Deficit from September 2014 to May 2015.

data was collected at hourly intervals throughout the study period (i.e. September 2014 to June 2015 in KBV).

2.4 Statistical analysis

Differences in the sapflux density (sapflow per unit sapwood area) were compared between the two cultivars using one way analysis of variance (t test) with $\alpha = 0.05$.

3 Results and discussion

The maximum daily solar radiation was 30.02 $MJ\,m^{-2}\,d^{-1}$ recorded on 17 December 2014 and the lowest was 6.9 $MJ\,m^{-2}\,d^{-1}$ measured on 2 November 2014. December 2014 had the highest average daily solar radiation of approx. 28.3 $MJ\,m^{-2}$. The maximum and minimum air temperatures were 37.3 and 1.4 °C reached in December 2014

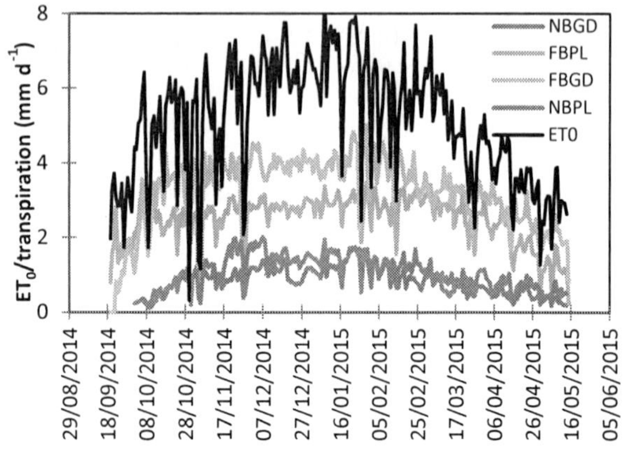

Figure 2. Water use of the young and mature apple trees and the reference evapotranspiration.

and June 2015, respectively. The maximum air temperature exceeded 30 °C in seven of the nine months. The relative humidity ranged from 99.9 % to as low as 5.4 % on hot dry days with the vapour pressure deficit of the air peaking at 3.1 kPa (Fig. 1). The reference evapotranspiration (ET_o), which is a measure of the atmospheric evaporative demand, was calculated as the evapotranspiration from a short grass reference using the modified Penman-Monteith equation according to Allen et al. (1998). The seasonal total reference evapotranspiration was 1205.4 mm. Transpiration for the full bearing "Golden Delicious" trees peaked at 5 $mm\,d^{-1}$, while the similar size trees of the Cripps' Pink cultivar peaked at 3.5 $mm\,d^{-1}$ at the end of January 2015. The non-bearing "Golden Delicious" and Cripps' Pink trees transpiration peaked at 1.7 and 2.02 $mm\,d^{-1}$ (Fig. 2) on 25 January 2015 and 6 December 2014, respectively. The seasonal total for the full-bearing "Golden Delicious" was 770 mm while for "Cripps' Pink" it was 587 mm. The non-bearing "Golden Delicious" trees transpired a seasonal total of 187 mm while "Cripps' Pink" transpired 238 mm for the season. Variations in transpiration rates in full-bearing orchards were a result of differences in canopy management practices between the "Golden Delicious" and "Cripps' Pink" orchards. "Cripps' Pink" canopies are kept open e.g. by pruning and spraying shoot growth retardants such as Regalis® to expose the fruit to solar radiation for anthocyanin synthesis to occur and to promote the development of the red fruit colour. The leaf area index (LAI) for the full bearing "Golden Delicious" was ~ 3.5 while for the "Cripps' Pink" it was ~ 2.8. For non-bearing orchards it was 0.8 for "Golden Delicious" and 1.0 for "Cripps' Pink". Canopy cover seems to be the main driver of water use, the bigger the canopy, the more water the tree uses. The daily total transpiration was linearly related to the daily solar radiation (Fig. 3a and b). However, the relationship between transpiration and VPD (Fig. 3c and d) was non-

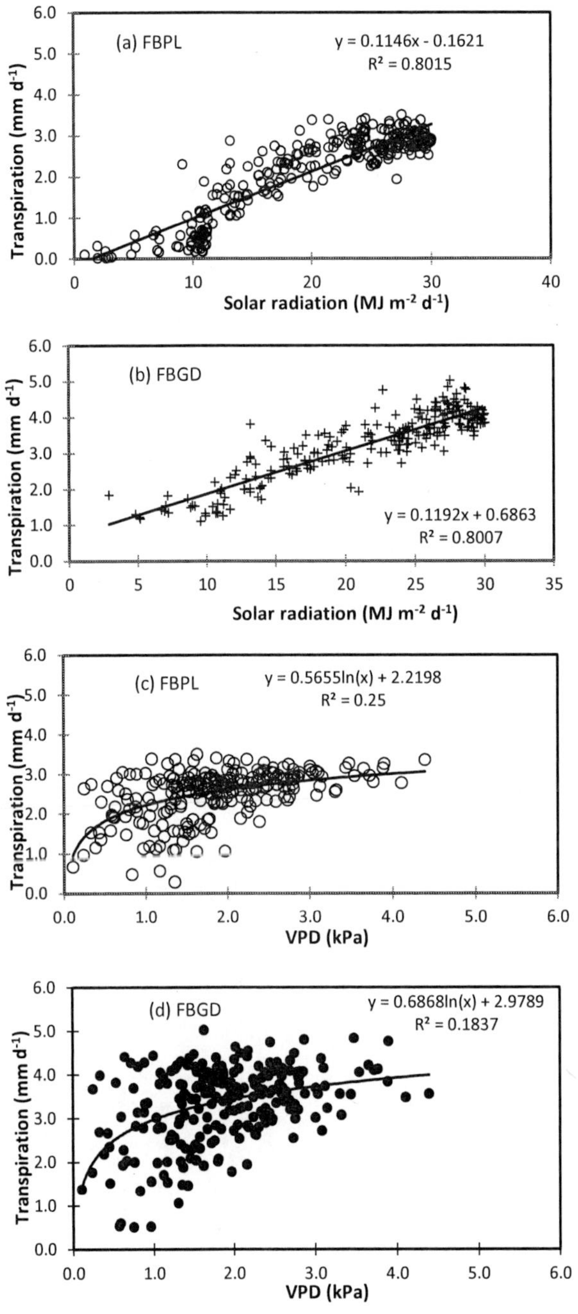

Figure 3. Transpiration response of full-bearing **(a)** "Cripps' Pink" to solar radiation, **(b)** "Golden Delicious" to solar radiation, **(c)** "Cripps' Pink" to VPD **(d)** "Golden Delicious" to VPD.

linear, suggesting that the VPD is the most limiting variable for apple trees transpiration.

4 Conclusions

This study provides first estimates of the water use characteristics of apple trees considering newly planted and mature trees. For mature trees, it appears that "Cripps' Pink" trees tend to use less water than "Golden Delicious". This is because they tend to have more open canopies due to management practices e.g. the spraying of chemicals such as Regalis® to reduce shoot growth. Dense foliage is not favourable for the development of the red colour on fruit in red and blushed cultivars and these canopy management practices appear to have water saving benefits. In non-bearing trees, the Cripps' Pink cultivar seems to be using more water than the "Golden Delicious". Transpiration was linearly related to the solar radiation, while VPD seemed to be a limiting variable for water use.

Data availability. The Water Research Commission (WRC) has all the rights to the data as they are the funders. Their policy supports that data be made public. However, this project is still ongoing, therefore no data is made available yet. When the project has been finalized, the data will be available at www.wrc.org.za.

Competing interests. The authors declare that they have no conflict of interest.

Special issue statement. This article is part of the special issue "Understanding spatio-temporal variability of water resources and the implications for IWRM in semi-arid eastern and southern Africa". It is a result of the IAHS Scientific Assembly 2017, Port Elizabeth, South Africa, 10–14 July 2017.

Acknowledgements. This project was initiated, supported and managed by the Water Research Commission (project no WRC K5 2398/4) and the South African Apple and Pear Producers' Association (SAAPPA). We are sincerely grateful for the financial support from these entities. Additional South African Parliamentary Grant funding through the Council for Scientific and Industrial Research (project number ECHS 043) and CSIR Young Researcher Establishment Fund (project ECHS 074) are also greatly acknowledged.

Edited by: Webster Gumindoga
Reviewed by: Timothy Dube and one anonymous referee

References

Allen, R. G., Pereira, L. S., Raes, D., and Smith, M.: Crop evapotranspiration, FAO irrigation and drainage paper 56, Food and Agriculture Organization, Rome, Italy, 1998.

Braun, P., Maurer, B., and Heinemann, G.: Scaling transpiration in apple orchards-meteorological versus plant based physiological measurements, Acta Hort., 537, 45–51, 2000.

Burgess, S. S. O., Adams, M. A., Turner, N. C., Beverly, C. R., Ong, C. K., Khan, A. A. H., and Bleby, T. M.: An improved heat pulse method to measure low and reverse rates of sap flow in woody plants, Tree Physiol., 21, 589–598, 2001.

Dragoni, D., Lakso, A. N., and Piccioni, R. M.: Transpiration of apple trees in a humid climate using heat pulse sap flow gauges calibrated with whole-canopy gas exchange chambers, Agr. Forest Meteorol., 130, 85–94, 2005.

Fereres, E., Goldhamer, D. A., and Sadras, V. O.: Yield responses to water of fruit trees and vines, UN Food and Agriculture Organisation, Rome Italy, 246–497, 2012.

Granier, A. R.: Evaluation of transpiration in a Douglas fir stand by means of sap flow measurements, Tree Physiol., 3, 309–320, 1987.

Gush, M. B. and Taylor, N. J.: The water use of selected fruit tree orchards (Volume 2): technical report on measurements and modelling, Water Research Commission report no. 1770/2/14, WRC, Pretoria, South Africa, 2014.

Midgley, S. J. E. and Lötze, E.: Climate change in the western cape of South Africa: trends, projections and implications for chill unit accumulation, Acta Hort., 903, 1127–1133, 2011.

Steppe, K., De Pauw, D. J. W., Doody, T. M., and Teskey, R. O.: A comparison of sap flux density using thermal dissipation, heat pulse velocity and heat field deformation methods, Agr. Forest Meteorol., 150, 1046–1056, 2010.

Swanson, R. H. and Whitfield, D. W. A.: A numerical analysis of heat pulse velocity theory and practice, J. Exp. Bot., 32, 221–239, 1981.

Volschenk T., De Villiers, J. F., and Beukes, O.: The selection and calibration of a model for irrigation scheduling of deciduous fruit orchards, WRC report 89, 2003.

Wullschleger, S. D. and King, W. K.: Radial variation in sap velocity as a function of stem diameter and sapwood thickness in yellow-poplar trees, Tree Physiol., 20, 511–518, 2000.

https://doi.org/10.5194/piahs-378-85-2018

Open Access
PIAHS

Near real time water quality monitoring of Chivero and Manyame lakes of Zimbabwe

Ronald Muchini[1], **Webster Gumindoga**[2], **Sydney Togarepi**[1], **Tarirai Pinias Masarira**[2], **and Timothy Dube**[3]

[1]University of Zimbabwe, Dept. of Geoinformatics and Surveying, Box MP 167, Mt Pleasant Harare, Zimbabwe
[2]University of Zimbabwe, Dept. of Civil Engineering, Box MP 167, Mt Pleasant, Harare, Zimbabwe
[3]Dept. of Earth Sciences, University of the Western Cape, Private Bag X17, Bellville, South Africa

Correspondence: Webster Gumindoga (wgumindoga@gmail.com)

Received: 16 November 2017 – Revised: 22 February 2018 – Accepted: 2 March 2018 – Published: 29 May 2018

Abstract. Zimbabwe's water resources are under pressure from both point and non-point sources of pollution hence the need for regular and synoptic assessment. In-situ and laboratory based methods of water quality monitoring are point based and do not provide a synoptic coverage of the lakes. This paper presents novel methods for retrieving water quality parameters in Chivero and Manyame lakes, Zimbabwe, from remotely sensed imagery. Remotely sensed derived water quality parameters are further validated using in-situ data. It also presents an application for automated retrieval of those parameters developed in VB6, as well as a web portal for disseminating the water quality information to relevant stakeholders. The web portal is developed, using Geoserver, open layers and HTML. Results show the spatial variation of water quality and an automated remote sensing and GIS system with a web front end to disseminate water quality information.

1 Introduction

Harare the capital city of Zimbabwe and its satellite towns (commonly known as Greater Harare) are seated in the Upper Manyame sub-catchment that supplies its water resources. Chivero and Manyame Lakes are the major sources of water for Greater Harare which are also situated downstream of the sub-catchment. It follows that any land use related activity in the Upper Manyame sub-catchment will affect the water status of the lakes. The Lakes are now rated among Zimbabwe and the world's most polluted inland water bodies (Chigonda, 2011; Chawira et al., 2013) and Chivero is rated among the world's most polluted lakes (Herald, 2012). The leading sources of pollution in these lakes have been identified by Kibena et al. (2013) as sewage (raw or partially treated) from sewer treatment plants as well as urban runoff. According to Buka et al. (2014), all the major sewage treating plants have been releasing their effluent into the local river system. Industries release their waste effluent into the river system. Leaching of nutrients from urban agricultural activities is also affecting the nutrient content of the lakes. Masere et al. (2012) and Kibena et al. (2013) supports the fact that the water quality of Upper Manyame has been impaired over the years.

A massive rise in urban population was realized over the past years (20–34 % between 1992 and 2002) and this hasled to the population overriding the urban utilities, hence the release of raw sewer into water bodies (Mapira, 2011). The deteriorating water quality is having adverse health effects on the populace, as well as leading to high water treating costs (Mlambo, 2013). The cholera outbreaks in Zimbabwe of 2002 and 2008 which affected about 11 735 lives with 484 deaths are attributed to poor sanitation and provision of polluted water to residents (Buka et al., 2014). Such deadly impacts of water pollution on humanity makes it of paramount importance to perform regular and synoptic water quality assessment and monitoring in the above-mentioned water bodies.

Water quality monitoring involves the retrieval of water quality parameters from the water for water management purposes (Dept. of Ecology State of Washington, 2012;

Published by Copernicus Publications on behalf of the International Association of Hydrological Sciences.

EMA, 2014). The current methods of water quality monitoring in Zimbabwe involves only the use of in-situ and laboratory based methods (EMA, 2012). These methods are costly, tedious, time consuming and do not provide complete spatial coverage of the water bodies and the water quality parameters of concern. This leaves the nation and the responsible authorities in particular with no option but to find and implement more efficient integrated water response strategies.

The applicability of remote sensing in retrieval of water quality variables, use of programming in automating GIS and remote sensing tasks and sharing of GIS products over the world wide web network allows for the development of an integrated and automated system for monitoring water quality and disseminating the results in near real time.

In this study we develop a near real-time water quality monitoring system for Chivero and Manyame lakes of Zimbabwe. This is achieved through identifying suitable algorithms that can determine water quality parameters and develop an application, which computes water quality parameters and present them on a web GIS.

2 Materials and methods

2.1 Study area

Chivero and Manyame Lakes are situated in the Upper Manyame sub-catchment of Zimbabwe which also contains the Greater Harare. The Surface area for Chivero and Manyame is about 26.3 and 81 km^2 (ZRA, 1998). The lakes are located at a lower altitude than most areas of the catchment including the city and all of its residential areas and industries. Rainfall received in the sub-catchment is nearly 798.6 $mm\,yr^{-1}$. Surface runoff especially during the wet season will always carry pollutants to the lakes. Lake Chivero has been hypertrophic since the 1960s and Manyame is now showing evidence of being hypertrophic as well (Magadza, 2008; Dlamini, 2012). The two lakes are currently Greater Harare's main source of water hence the need for them to be closely monitored.

2.2 Data availability

Remote sensing and field based (in-situ) data were applied. Field based water quality data was obtained from in situ measurements for Chivero at three selected sampling points by Dlamini (2012). Landsat 5–7 Thematic Mapper (TM) and Enhanced Thematic Mapper (ETM+), freely available from the United States Geological Survey (USGS) website were used for this research. The imagery has a 30 m and 16 day spatial and temporal resolution respectively. Only imagery with less than 30 % cloud cover was selected for use.

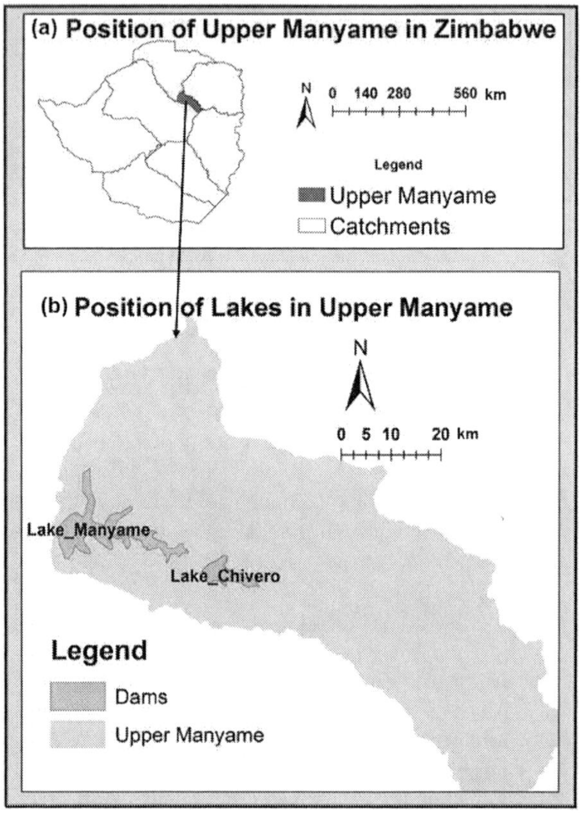

Figure 1. Map of study area (first: Upper Manyame in Zimbabwe, second: Lakes Chivero and Manyame in Upper Manyame).

2.3 Determination of water quality parameters

Landsat Imagery was downloaded for April and September (1986, 1990, 2000, 2005, 2012 and 2015) from the USGS glovis website. Image preprocessing was done through the conversion from Digital Numbers (DN) to spectral radiance by implementing the Chander et al. (2009) algorithm using the ENvironment for Visualizing Images (ENVI) software. The water quality parameters were computed in ILWIS. The computed raster maps were resampled to a common georefrence and the lake are extracted. The water quality parameters were validated using the point to pixel method (Chander et al., 2009).

The authors used formulae in Shafique et al. (2002) and Dlamini (2012) as shown in Table 1. Several formulas applied before in tropical lakes have been tested and it is the formula from the above authors that yielded better correlation with ground based measurements and thus has been adopted in this study.

2.4 Application for automated retrieval of water quality parameters

ILWIS software was used for automating retrieval.Making use of of ILWIS's DDE server capabilities and VB6's DDE client capabilities, the automation of water quality parameters computation was achieved using VB6. A client program to ILWIS was programmed in VB6 also using ILWIS scripting commands (ILWIS, 2013; Microsoft, 2015). The application provides navigation tools for the user to navigate through the host computer and and to select the folder containing the images and click a command button so that the automated computations begin.

2.5 Dissemination of water quality data via the web

Geoserver was installed and the necessary set up done. The web interface and client to the data was programmed using HTML, JavaScript and PHP scripting languages, as well as the OpenLayers mapping library. It was designed in such a way to allow the user to enter date and parameter of interest, and then the program would retrieve the specified parameter at the specified date and present it as a map.

3 Results and discussion

3.1 The spatial variation of water quality parameters

The research presents four successfully determined water quality parameters; Chlorophyll *a*, turbidity, totals Phosphorus and Total Suspended Matter. Figures 2–5 show the spatial and temporal variations of the water quality parameters. The parameter values from 0 to a maximum are represented on the maps with colour variations from blue to red with interpolation in between.

3.1.1 Chlorophyll *a* (Chl *a*)

The minimum value for Chl *a* was 0.00 mug$\,$L^{-1} for all years whilst the maximum varied between 0.15 and 0.23 mug$\,$L^{-1} without any particular order of change. However, an increase in the amount of Chl *a* can be noted by the increase in the redness of the map. There is a notable increase in the pollution with respect to Chl *a* over time with the 2005 being an exemption in the trend due to its extremely low values as shown by the blue color intensity. Figure 2 shows the trends explained here. The values 0.15 to 0.23 mug$\,$L^{-1} show a hypertrophic lake and therefore agree with Nhapi (2008) who reiterated that lake Chivero has been eutrophic since 1960 when the first algal blooms appeared. Dlamini (2012) also got values 0.101 ± 0.128 mug$\,$L^{-1} which basically falls in the same range as given in this work.

3.1.2 Turbidity (Tb)

There is a general increase in the pollution levels with respect to turbidity as it is evident on the diagrams by the increase in the intensity of red on the maps. The year 1995 is noted as an exception in the trend. Generally, the minimum value of the parameter increased from 1.2 NTU in 1986 to 3.6 NTU in 2015, whilst the maximum parameter value increased from 3.2 to 5.8 NTU throughout the study period. The spatial and temporal variations of turbidity are shown on Fig. 3. Turbidity values of less than 5 NTU are accepted by the Zimbabwean effluent standards. So, the water quality changed from an acceptable status of 3.2 to 5.8 NTU a status beyond the accepted.

3.1.3 Total Phosphorus (TP)

The minimum values of total phosphorus varied from 0.00 to 0.91 mg$\,$L^{-1} from 1986 to 2015. In the same vein, the maximum total phosphorus varied from 0.04 to 6.82 mg$\,$L^{-1}. This variation shows that the water quality (with respect to TP) has been deteriorating with time from 1986. This could probably be due to increasing Harare population and therefore increased use of fertilizers in urban agriculture which would leach into the water system. Generally, the mean values shows a hypertrophic lake with respect to TP and this is in sync with the work of Nhapi (2008) who postulated that the lakes were eutrophic since 1960. Dlamini (2012) also found the lake hypertrophic at 1.17 mg$\,$L^{-1}. Figure 4 shows the spatial and temporal variations of TP.

3.1.4 Total Suspended Matter (TSM)

Figure 5 shows the spatial and temporal variation of TSM. The variations of water pollution status with respect to TSM were actually retrogressive in that from 1986 to 2015 the maximum and minimum values varied from 3.25 to 1.77 mug$\,$L^{-1} and 0.39 to 0.37 mug$\,$L^{-1} respectively. However, the retrogression was not gradual, an increase is noted from 1995 to 2005 on the maximum values as well as a slight increase to 2015. The decrease is probably explained by the growth of weeds and other forms of vegetation along the edges of the lakes such that the suspended solids heading for the water are trapped in the weeds before reaching the water itself.

3.1.5 Summary of all retrieved parameters

Table 2 is a summary of the water quality parameters, showing the maximum and minimum of each parameter from 1986 to 2015. The parameters show that the lake is polluted and pollution generally increased with time.

Table 1. Algorithms for retrieval of water quality parameters from Landsat images.

Parameter	Algorithm i.d	Algorithm	Units
Chlorophyl *a*	3.4.1...	$\text{Chl } a = (486 \cdot \text{NDVI} + 44) \cdot -0.001$ $\text{NDVI} = (\text{B4} - \text{B3})/(\text{B4} + \text{B3})$	$\mu g\,L^{-1}$
Turbidity	3.4.2...	$\text{Tb} = 5.0819 \cdot \left(\frac{\text{B1}}{\text{B3}}\right)^{-1.0125}$	NTU
Total Phosphorus	3.4.3...	$\text{Ln(TP)} = 4.334 - 4.594\left(\frac{\text{B1}}{\text{B2}}\right) + 1.103\left(\frac{\text{B4}}{\text{B7}}\right)$	$mg\,L^{-1}$
Total Suspended Matter	3.4.4...	$\text{TSM} = 3.39 \cdot \left(\frac{\text{B2}}{\text{B3}}\right) - 2.6$	$\mu g\,L^{-1}$

(Shafique et al., 2002; Dlamini, 2012)
NB B1–B7 refers to Band 1 to Band 7 of Landsat TM/ETM+ in all equations in Table 1.

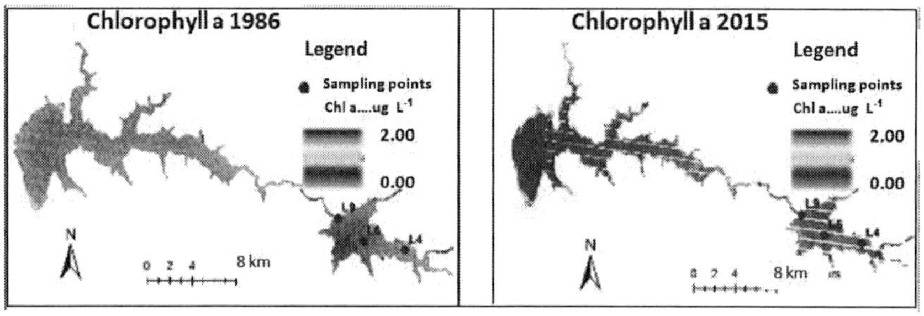

Figure 2. Spatial variations of Chlorophyll *a* 1986 and 2015.

Table 2. Summary of the estimated water quality variables.

Date	Parameter	Min	Mean	Max	Units
31 Aug 1986	Chl *a*	0.0000	0.040	0.2184	$mug\,L^{-1}$
	Turbidity	1.2486	1.590	3.2739	NTU
	TP	0.0000	1.590	22 726	$mg\,L^{-1}$
	TSM	0.0000	1.580	20.246	$mug\,L^{-1}$
10 Aug 1990	Chl *a*	0.0000	0.010	0.1698	$mug\,L^{-1}$
	Turbidity	0.9764	1.390	3.5554	NTU
	TP	0.0001	0.010	13.527	$mg\,L^{-1}$
	TSM	0.0000	1.940	4.1800	$mug\,L^{-1}$
8 Aug 1995	Chl *a*	0.0000	0.001	0.2330	$mug\,L^{-1}$
	Turbidity	1.2756	2.670	4.6672	NTU
	TP	0.0005	0.551	411 319	$mg\,L^{-1}$
	TSM	0.0000	0.520	2.4850	$mug\,L^{-1}$
13 Aug 2000	Chl *a*	0.0000	0.040	0.1552	$mug\,L^{-1}$
	Turbidity	1.9733	3.310	5.9002	NTU
	TP	0.0614	1.510	494.10	$mg\,L^{-1}$
	TSM	0.0000	1.370	2.7932	$mug\,L^{-1}$
4 Sep 2005	Chl *a*	0.0000	0.002	0.1164	$mug\,L^{-1}$
	Turbidity	1.5518	2.050	3.6652	NTU
	TP	0.0011	0.020	6.4145	$mg\,L^{-1}$
	TSM	0.0000	0.920	1.7145	$mug\,L^{-1}$
19 Jan 2015	Chl *a*	0.0000	0.100	0.2135	$mug\,L^{-1}$
	Turbidity	3. 6161	4.400	5.8373	NTU
	TP	0.9099	4.120	187.82	$mg\,L^{-1}$
	TSM	0.3704	1.360	1.7673	$mug\,L^{-1}$

Table 3. Validation of Chlorophyll *a*.

Sampling Point	Remotely sensed Chl *a* ($mug\,L^{-1}$)	In-situ measured Chl *a* ($mug\,L^{-1}$)
L4	0.180	0.201
L6	0.140	0.160
L9	0.126	0.144

3.1.6 Validation of retrieved parameters

Using data obtained from Dlamini (2012) and February 2012 Landsat imagery, the remotely sensed parameters were compared to in-situ derived data from three lake sampling points which are L4, L6, and L9.

Chlorophyll *a*

The remotely sensed remote sensing values for the four sites show a deviation of 0.021, 0.02 and 0.018 $mug\,L^{-1}$ for the sites L4, L6, and L9 respectively. So, on average, the remotely sensed data has about -0.02 $mug\,L^{-1}$ deviation from the in situ-derived values. Table 3 shows the relationship between remotely sensed Chl *a* and In-situ derived Chl *a*.

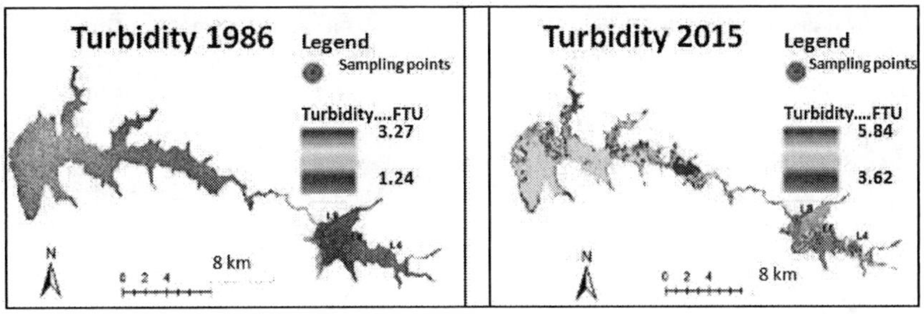

Figure 3. Spatial variations of Turbidity 1986 and 2015.

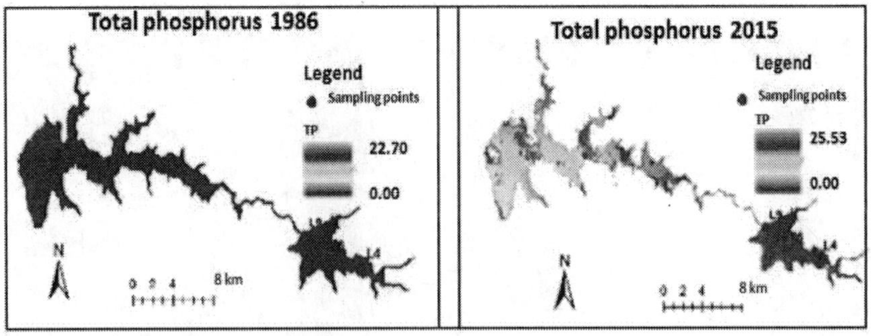

Figure 4. Spatial variations of Total Phosphorus 1986 and 2015.

Table 4. Validation of total phosphorus.

Sampling Point	Remotely sensed TP ($mg L^{-1}$)	In-situ measured TP ($mg L^{-1}$)
L4	0.337	0.880
L6	0.972	1.500
L9	0.600	1.130

Total phosphorus

The remotely sensed values of total phosphorus show deviations of 0.543, 0.528 and 0.530 for L4, L6 and L9 site respectively. Hence an average of −0.53 deviation from the in-situ derived is observed for total phosphorus. Table 4 shows in situ derived TP and remotely sensed TP.

3.2 Automation of water quality parameters' retrieval

The determination of parameters was successfully automated, from the stage of importing them up to the finished water quality maps. This was achieved using the vb6 application with the interface shown on Fig. 6.It is the back-end application which the personnel responsible for water quality data dissemination will be using to retrieve water quality parameters from satellite imagery. The interface allows the user to navigate to a directory where a satellite image is stored and by clicking the locate button and then the proceed button that comes on top of the locate button in about 10 s after clicking the locate button. Kabo-bah (2010) also developed a water quality modelling scheme which was integrated with ILWIS. However, his system was installed as a plug in in ILWIS and therefore requiring the users to have knowledge of the software in to be able to manipulate it.

3.3 The web dissemination of water quality data

The dissemination of water quality data to the web was achieved using Geoserver accessing imagery in a file folder and OpenLayers. The client main webpage provides the user with three dropdown menus, the other one to select year, another month and the other one to select the parameter which the user wants to view. There is also a submit button which the user uses to submit their query to the host and retrieve the desired water quality parameter.

The buttons captioned "about us", "database"and "contact us" lead to the respective web pages as labelled.

Mouse location on the lakes is also shown by coordinates in WGS84 UTM, on the bottom right end of the map section. Steissberg et al. (2010) developed a web accessible repository similar to the one presented in this paper for Lake Tahoe, South Nevada. However, theirs allowed users to make downloads of the water quality products, functionality not provided for here. Another web GIS system comparable to this one was developed by Li et al. (2006), however this one was for exploration and visualisation of ice in Arctic waters.

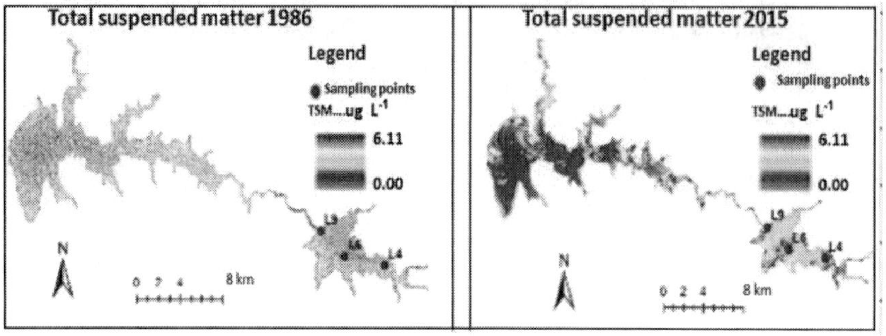

Figure 5. Spatial variations of Total Suspended Matter 1986 and 2015.

Figure 6. Interface of the backend application for automated remote sensing and GIS.

Stakeholders that could benefit from this system include the Environmental management Agency (EMA), Zimbabwe National Water Authority (ZINWA), the Upper Manyame sub-Catchment Council (UMSCC) and those responsible for pollution. ZINWA for decision making as they mainly deal with surface water abstraction. EMA and local authorities can use this to raise awareness to the public on how much their activities are affecting the environment for effective clean up and environmental campaigns.

4 Conclusions

Three conclusions were drawn.

1. Four water quality parameters were successfully determined that is Chlorophyll a, Turbidity, Total Phosphorus and Total Suspended Matter. These parameters are among the key indicators of water pollution and quality. Water quality of Chivero and Manyame lakes has been deteriorating that is specifically with respect to the determined water quality parameters in the study period.

2. The retrieval of water quality parameters can be automated by combining computer programming and remote sensing tools. In this work the determination of all the above-mentioned parameters were automated and the results from the automated process were found to be tallying with those obtained when the processes were executed manually. Water quality data can be made accessible to the public easily and on a near real time basis using web GIS technology.

3. There is therefore a need to apply remote sensing and GIS tools in water quality monitoring in order to reduce the costs of monitoring. This will also provide information of the spatial variations of the key water parameters and do away with the statistical assumptions of sampling. Spatial variations of water quality parameters can also put water treatment organizations in a better position by knowing lake areas for water withdrawal from at any given time and reduce treatment costs. The water quality parameter computing application eliminates human error in processing.

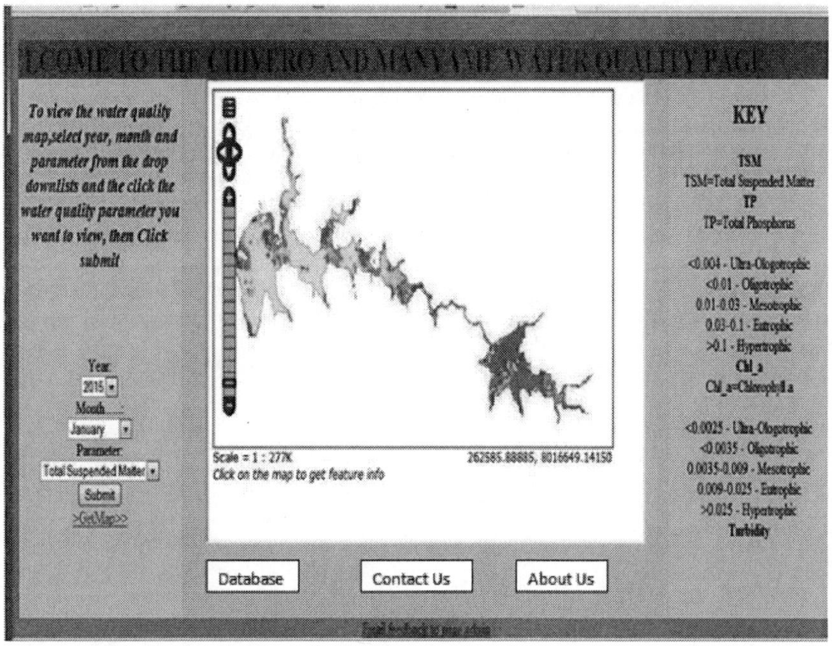

Figure 7. The web accessible interface to water quality data.

Data availability. Data may be accessed by emailing our first author.

Competing interests. The authors declare that they have no conflict of interest.

Special issue statement. This article is part of the special issue "Understanding spatio-temporal variability of water resources and the implications for IWRM in semi-arid eastern and southern Africa". It is a result of the IAHS Scientific Assembly 2017, Port Elizabeth, South Africa, 10–14 July 2017.

Edited by: Jean-Marie Kileshye-Onema
Reviewed by: Colleta Tundu and Nyaradzayi Mawango

References

Buka, H., Linhoss, A., and Pote, J.: An Integrated Watershed Approach to Water and Sanitation Hygiene priorities through a Narrative Review of Lake Chivero, Zimbabwe, Starkville, 2014.

Chander, G., Markham, B. L., and Helder, D. L.: Summary of Current Radiometric Calibration Coefficients for, Remote Sens. Environ., 1, 1–24, 2009.

Chawira, M., Dube, T., and Gumindoga, W.: Remote sensing based water quality monitoring in Chivero and Manyame lakes of Zimbabwe, Phys. Chem. Earth, 66, 38–44, https://doi.org/10.1016/j.pce.2013.09.003, 2013.

Chigonda, T.: Thirst in the midst of the Twin lakes: A quest for understanding Norton's ironical water woes, Journal of Sustainable Development in Africa, 13, 295–303, 2011.

Dept. of Ecology State of Washington: Water Quality Program Policy, USA, 2012.

Dlamini, S.: Evaluation Of The Water Quality Statusin Lake Chivero, Zimbabwe, University of Zimbabwe, 2012.

EMA.: Environmental Management Agency – Water testing services, available at: http://www.ema.co.zw/index.php/2014-06-1203-49-33/2014-06-12-03-50-38/ema-services/50-water-testing-services.html (last access: 22 April 2015), 2012.

EMA.: Ambient water quality monitoring, 2014, available at: http://www.herald.co.zw/ambient-water-quality-monitoring/ (last access: 20 March 2015), 2014.

Herald: The Herald, Lake Chivero Ranks High in pollution, 18 April, Lake Chivero Ranks High in pollution, available at: https://www.herald.co.zw/lake-chivero-ranks-high-in-pollution/ (last access: 10 March 2018), 2012.

ILWIS.: Apendices ILWIS as a DDE server(Adavanced). Available at: http://spatial-analyst.net/ILWIS/htm/ilwis/appendices_ilwis_as_dde_server.htm (last access: 23 April 2015), 2013.

Kabo-bah, A.: Developing water quality modelling scheme in ILWIS Open for the Dinkel River in support of climate change and adaptation studies Developing water quality modelling scheme in ILWIS Open for the Dinkel River in support of climate change and adaptation studi, Twente, Enschede, 2010.

Kibena, J., Nhapi, I., and Gumindoga, W.: Assessing the relationship between water quality parameters and changes in landuse patterns in the Upper Manyame River, Zimbabwe, J. Phys. Chem. Earth, 67–60, 153–163, https://doi.org/10.1016/j.pce.2013.09.017, 2013.

Li, S., Xiong, C., and Ou, Z.: Development Of Web-Based Ice Information System, Ou 2002, Canada, 2006.

Magadza, C. H.: Management of Eutrophication in Lake Chivero?, Success and failures?: A Case Study, Harare, 2008.

Mapira, J.: River pollution in the City of Mutare (Zimbabwe) and Its Implications For Sustainable Development, Journal of Sustainable Development in Africa, 13, 181–194, 2011.

Masere, T. P., Munodawafa, A., and Chitata, T.: Assessment of human impact on water quality along Manyame River, Int. J. Develop. Sustainability, 1, 754–765, 2012.

Microsoft: About Dynamic Data Exchange, available at: https://msdn.microsoft.com/en-us/library/windows/desktop/ms648774(v=vs.85).aspx (last access: 23 March 2018), 2015.

Mlambo, M.: Zimbabwe's river system heavily polluted, available at: www.thestandard.co.zw/2013/04/07/zimbabwes-river-system-heavily-polluted/ (last access: 17 February 2015), 2013.

Nhapi, I.: Inventory of water management practices in Harare, Zimbabwe, Water Environ. J., 22, 54–63, https://doi.org/10.1111/j.1747-6593.2007.00084.x, 2008.

Shafique, N. A., Fulk, F., Autrey, B. C., Flotemersch, J.: Hyperspectral Remote Sensing of Water Quality Parameters for Large Rivers in the Ohio River Basin, Environmental Protection Agency, USA, 2002.

Steissberg, T., Schladow, G., and Hook, S.: Monitoring past, present, and future water quality using remote sensing, Tahoe environmental research . . . , Southern Nevada, 2010.

ZRA: Water Consumption and Effluent from Food and Agricultural Sector including Fisheries and Livestock, Danida, 1998.

https://doi.org/10.5194/piahs-378-93-2018

Open Access
PIAHS

Isotopic composition and elemental concentrations in groundwater in the Kuiseb Basin and the Cuvelai-Etosha Basin, Namibia

Nnenesi A. Kgabi[1], Eliot Atekwana[2], Johanna Ithindi[1], Martha Uugwanga[1], Kay Knoeller[3], Lebogang Motsei[4], Manny Mathuthu[5], Gideon Kalumbu[1], Hilma R. Amwele[6], and Rian Uusizi[1]

[1]Department of Civil and Environmental Engineering, Namibia University of Science and Technology, Windhoek, Namibia
[2]Department of Geological Sciences, University of Delaware, Newark, Delaware, USA
[3]Department of Catchment Hydrology, Helmholtz Centre for Environmental Research, Leipzig, Germany
[4]School of Agriculture, North-West University, Mafikeng, South Africa
[5]Centre for Applied Radiation Science and Technology, North-West University, Mafikeng, South Africa
[6]Department of Agriculture & Natural Resources Sciences, Namibia University of Science and Technology, Windhoek, Namibia

Correspondence: Nnenesi A. Kgabi (nkgabi@nust.na)

Received: 11 November 2017 – Revised: 25 February 2018 – Accepted: 5 March 2018 – Published: 29 May 2018

Abstract. We assessed environmental tracers in groundwater in two contrasting basins in Namibia; the Kuiseb Basin, which is a predominantly dry area and the Cuvelai-Etosha Basin, which is prone to alternating floods and droughts. We aimed to determine why the quality of groundwater was different in these two basins which occur in an arid environment. We analysed groundwater and surface water for the stable isotope ratios of hydrogen (δ^2H) and oxygen (δ^{18}O) by cavity ring-down spectroscopy and metals by inductively coupled plasma mass spectrometry. The δ^2H and δ^{18}O of surface water in the Cuvelai-Etosha Basin plot on an evaporation trend below the global meteoric water line (GMWL) and the local meteoric water line (LMWL). The δ^2H and δ^{18}O of some groundwater samples in the Cuvelai-Etosha Basin also plot on the evaporation trend, indicating recharge by evaporated rain or evaporated surface water. In contrast, the δ^2H and δ^{18}O of groundwater samples in the Kuiseb Basin plot mostly along the GMWL and the LMWL, indicating direct recharge from unevaporated rain or unevaporated surface water. Fifty percent of groundwater samples in the Cuvelai-Etosha Basin was potable (salinity <1 ppt) compared to 79 % in the Kuiseb Basin. The high salinity in the groundwater of the Cuvelai-Etosha Basin does not appear to be caused by evaporation of water (evapo-concentration) on surface prior to groundwater recharge, but rather by the weathering of the Kalahari sediments. The low salinity in the Kuiseb Basin derives from rapid recharge of groundwater by unevaporated rain and limited weathering of the crystalline rocks. The order of abundance of cations in the Kuiseb Basin is Na > K > Ca > Mg vs. Na > Mg > Ca > K for the Cuvelai-Etosha Basin. For metals in the Kuiseb Basin the order of abundance is Fe > Al > V > As > Zn vs. Al > Fe > V > As > Zn for the Cuvelai-Etosha Basin. The relative abundance of cations and metals are attributed to the differences in geology of the basins and the extent of water-rock interaction. Our results show that the quality of groundwater in Cuvelai-Etosha Basin and Kuiseb Basin which vary in the extent of aridity, is controlled by the extent of water-rock interaction at the surface and in the groundwater aquifer.

Published by Copernicus Publications on behalf of the International Association of Hydrological Sciences.

1 Introduction

Water scarcity is one of the key limiting factors to sustainable economic, industrial, social and environmental development. In addition to scarcity, quality is a limiting factor for the sustainable use of water for domestic, agricultural and industry activities. For example, salinization which impairs the use of water is a global problem. Moreover, salinization is more severe in arid regions, where groundwater is the primary source of water (IAEA, 2006).

In arid environments, precipitation and groundwater recharge are extremely variable (Lewis and Walker, 2002). The spatial variability of rain recharge along permanent and intermittent river reaches and the variable water-rock interaction and groundwater mixing add complexity to investigating the origin and chemical evolution of groundwater. The source(s) of water and the physical and chemical evolution can be assessed by environmental tracers such as isotopes and the types of chemical constituents (e.g., major and trace elements) (Elliot, 2014). For example, the stable hydrogen (δ^2H) and oxygen (δ^{18}O) isotopic composition of groundwater depends on that of precipitation. However, the climatic conditions of the recharge environment and other local factors such as topography and the recharge process can cause variation in the δ^2H and δ^{18}O which can be traced in groundwater (Fritz et al., 1987; Clark and Fritz, 1997).

Namibia, located in an arid climate is becoming more dependent on groundwater, especially during droughts. The quality of the groundwater obtained from different groundwater basins are highly variable. In general, groundwater conditions are unfavourable due to limited availability, little and unreliable recharge, low borehole yields, poor groundwater quality and high risks of contamination (Christelis and Struckmeier, 2011). It is therefore important to assess groundwater in order to understand the factors that cause variability in the quality. In this study, we use environmental tracers to assess the nature of recharge and control of the quality of groundwater in the Kuiseb Basin and Cuvelai-Etosha Basin. Potential evapotranspiration in the Kuiseb Basin (785–1241 mm yr^{-1}) and Cuvelai-Etosha Basin (1880–2173 mm yr^{-1}) are variable (Kaseke et al., 2016). On an annual basis, the rainfall frequencies and intensity vary significantly in the Kuiseb Basin (8–255 mm yr^{-1}) and Cuvelai-Etosha Basin (410–690 mm yr^{-1}) (e.g., Kaseke et al., 2016), which determines how rain is affected prior to recharge of groundwater. In the Kuiseb Basin, there is limited surface ponding and runoff and the basin is predominantly dry. In contrast, the Cuvelai-Etosha Basin is prone to alternating floods and droughts and water ponds periodically on the surface and there is intermittent river flow during flood periods. In the Kuiseb Basin, the quality of groundwater although portable is highly variable across the basin while in the Cuvelai-Etosha Basin, groundwater quality is mostly poor due to high salinity (Falke, 2008).

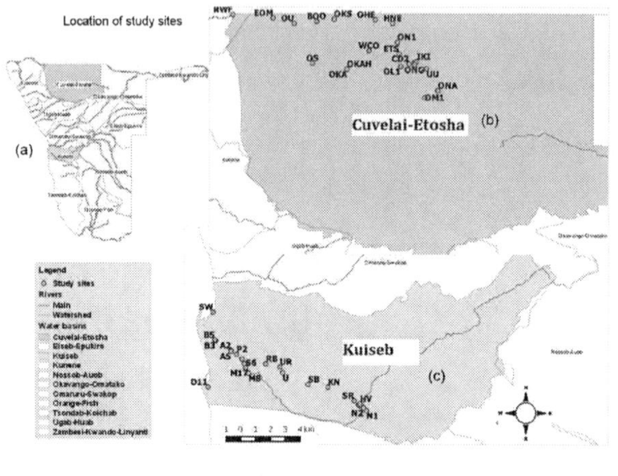

Figure 1. Map of Namibia showing the drainage basins **(a)** and sampling station for the Cuvelai-Etosha Basin **(b)** and the Kuiseb Basin **(c)**.

2 Study Area

The Kuiseb Basin is located in the western part of Namibia (Fig. 1) within the Namib Desert that stretches parallel to the coast.

The Kuiseb Basin is underlain by a crystalline basement of metamorphic sedimentary rocks (schist, quartzite or marble) with granitic intrusions (Miller, 1980; Schneider, 2004) and sediments derived from these rocks. The Kuiseb Basin is divided into an upper watershed which consists of highlands and commercial farms, a middle watershed which has the Namib-Naukluft Park and small-scale farms (predominantly Topnaar communities) and a lower watershed close to the coast encompassing the Walvis Bay and surrounding area. The Basin is drained by the Kuiseb River (Falke, 2008) and flows only after extremely heavy rains such as in 2011 (Mendelson et al., 2013; Gardiner et al., 2006). The two aquifers in the Kuiseb Basin are located at Swartbank and Rooibank in the lower portion of the basin. Groundwater in the Kuiseb Basin is recharged exclusively from rainfall and runoff (Heyns and van Vuuren, 2009).

The Cuvelai-Etosha Basin is located in the north-central part of Namibia. The Cuvelai-Etosha Basin is covered by limestone and dolomite rocks and Kalahari sediments (unconsolidated to semi-consolidated sands, gravels and silts and calcrete) (Mendelson et al., 2013). Groundwater in the basin is mainly abstracted from the Ohangwena Kalahari Aquifer and discontinuous perched aquifers by means of boreholes. Wells are used to supply water, especially to isolated villages in the basin (Hamukoto et al., 2017). The groundwater is saline except for the Tsumeb Karst.

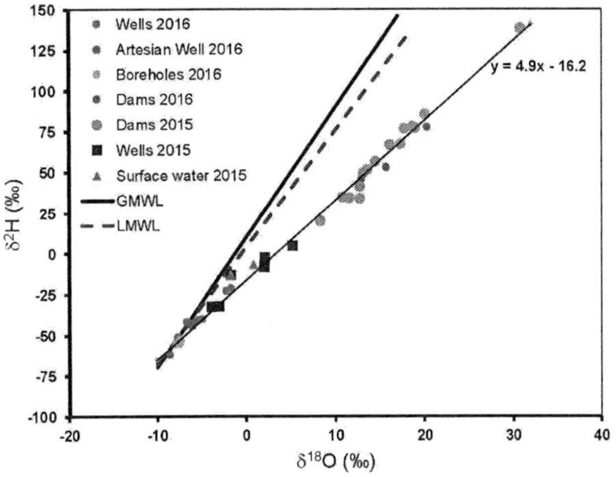

Figure 2. Plot of the stable oxygen isotopes (δ^{18}O) vs. hydrogen (δ^{2}H) for water samples from the Cuvelai-Etosha Basin. GMWL is the global meteoric water line and LMWL is the local meteoric water line.

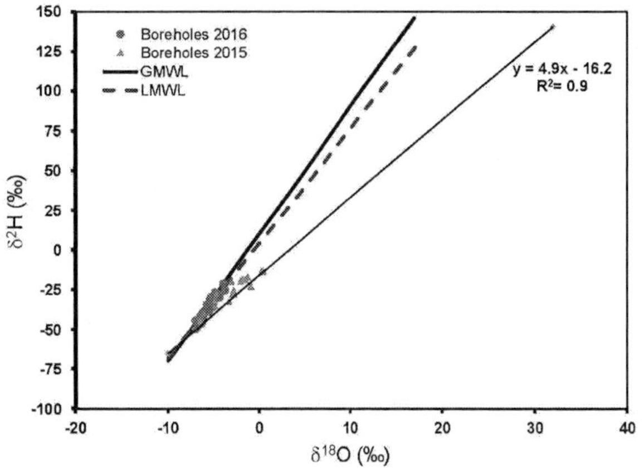

Figure 3. Plot of the stable oxygen isotopes (δ^{18}O) vs. hydrogen (δ^{2}H) for water samples from the Kuiseb Basin. GMWL is the global meteoric water line and LMWL is the local meteoric water line.

3 Methods

We collected and analysed water samples from groundwater and surface water in August to November of 2015 and in August to September 2016. Fifty-six sampling points were investigated; 25 sampling points from the Cuvelai-Etosha Basin and 31 from the Kuiseb Basin.

Groundwater samples were collected from wells (hand dug) and boreholes. The groundwater was pumped to the surface for about 5 min to obtain water from the formation. Surface water was sampled by the grab techniques along rivers and dams. An artesian groundwater was sampled where it issued to the surface. All water for cations and anions were collected in polyethylene bottles and stored in cooler boxes. The samples were filtered in the laboratory and refrigerated until analyses. Water for stable isotopes were collected in glass vials, kept cool in the field and refrigerated in the laboratory until analyses.

Water temperature was measured onsite and the total dissolved solids (TDS), salinity and electrical conductivity were measured either onsite using a portable HQ40d Dual Input Multi-parameter instrument or in the laboratory using microelectrodes. Elemental concentrations were determined by EPA Method 200.8 using an inductively coupled plasma mass spectrometry (NexION 300D/350D) as described by Pruszkowski and Bosnak (2014). Water analysed for δ^{2}H and δ^{18}O was done by Cavity Ring-Down Spectrometry (Picarro L2120-i) and are reported in the standard delta (δ) notation in per mil (‰) relative to the VSMOW for the H and O isotopes.

4 Results and Discussion

4.1 Isotope Ratios

The δ^{2}H and δ^{18}O from the Cuvelai-Etosha Basin and the GMWL (Craig, 1961) and LMWL ($\delta^{2}\text{H} = 7.2\,\delta^{18}\text{O} + 4.1$; Turewicz, 2013) are shown in Fig. 2.

The δ^{2}H and δ^{18}O of surface waters are enriched and plot along a line below the GMWL and the LMWL, reflecting high evaporation. Some groundwater samples lie along the evaporation line and were subject to evaporation prior to groundwater recharge (Prada et al., 2010).

The δ^{2}H and δ^{18}O from the Kuiseb Basin are shown in Fig. 3. Also shown are the GMWL and the LMWL. Compared to the Cuvelai-Etosha Basin, groundwater in the Kuiseb Basin were more depleted and the δ^{2}H and δ^{18}O plot mostly along the GMWL and the LMWL and at the lower end of the evaporation line.

In principle, the δ^{2}H and δ^{18}O of groundwater should be related to that of the local precipitation. The δ^{2}H and δ^{18}O of groundwater co-vary, and on a global basis lie on the GMWL or on the LMWL (e.g., Clark and Fritz, 1997). The GMWL and the LMWL have been used in various studies as a convenient reference for tracing the origin and evolution of water in the hydrologic cycle (McGuire and McDonnel, 2007). When the rain or surface water is subjected to evaporation, the co-variation between the δ^{2}H and δ^{18}O deviate and lie below the GMWL and LMWL; the extent of the deviation depends on the aridity.

In a study of the distribution of δ^{2}H and δ^{18}O in rain across Namibia, Kaseke et al. (2016) modelled heavier δ^{2}H and δ^{18}O for rains in the Kuiseb Basin and more depleted values for rains in the Cuvelai-Etosha Basin. Yet, the δ^{2}H and δ^{18}O in groundwater in the Cuvelai-Etosha Basin (Fig. 2) are more enriched compared to those of groundwater the Kuiseb

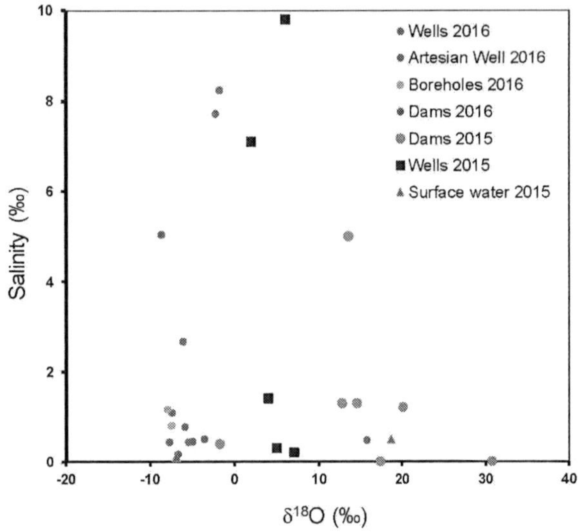

Figure 4. Plot of the stable oxygen isotopes ($\delta^{18}O$) vs. salinity for water samples from the Cuvelai-Etosha Basin.

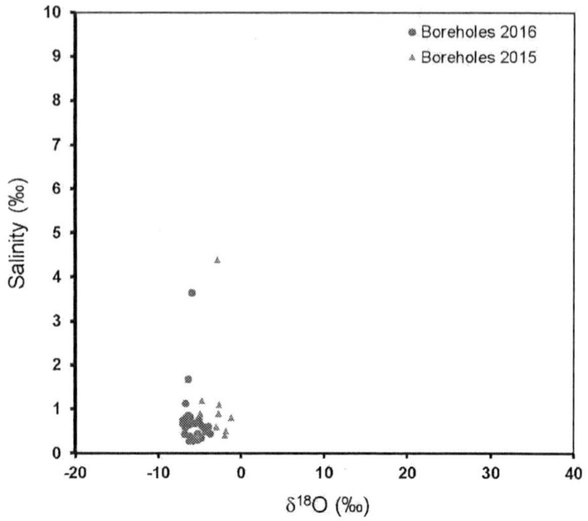

Figure 5. Plot of the stable oxygen isotopes ($\delta^{18}O$) vs. salinity for water samples from the Kuiseb Basin.

Basin (Fig. 3). The δ^2H and $\delta^{18}O$ in some groundwater in the Cuvelai-Etosha Basin indicates extensive evaporation before groundwater recharge. This is because rainwater is trapped on the surface during flooding by muddy sediments deposited by floods which reduce rapid infiltration (Mendelson et al., 2013). This extensive evaporation of rain prior to groundwater recharge appears to be limited in the Kuiseb Basin (Fig. 3).

4.2 Groundwater Quality

The salinity of groundwater can be used as a proxy for the ionic content (similar to the electrical conductivity and TDS) and reflect processes that increase the solute content in water. The salinity levels in groundwater in Cuvelai-Etosha Basin are generally high (e.g., Fig. 4). About 50 % of the groundwater have salinity lower than marginal (< 1 ‰) quality and 50 % have salinity that makes the groundwater brackish to saline (1–10 ‰) quality.

The salinity levels measured for the Kuiseb Basin are much lower (e.g., Fig. 5) compared to groundwater in the Cuvelia-Etosha Basin. About 79 % of groundwater in the Kuiseb have salinity lower than marginal (< 1 ‰) quality and 21 % have salinity that makes the groundwater brackish to saline (1–10 ‰) quality. The high salinity levels in the Cuvelai-Etosha Basin may be due to the weathering of abundant limestone, dolomite and calcrete (Msangi, 2014). Also, high evaporation rates indicated by the enriched δ^2H and $\delta^{18}O$ of some groundwater samples point to evapo-concentration of solutes (Harbeck, 1955). Therefore, the lower salinity levels in the Kuiseb Basin may thus reflect limited weathering of the crystalline basement rocks and lower evapo-concentration indicated by more depleted δ^2H and $\delta^{18}O$ in groundwater.

The effects of rainfall and evaporation vs. water-rock interaction on groundwater quality can be assessed by examining the relationship between the $\delta^{18}O$ (or δ^2H) and solute concentrations (Winston and Criss, 2002) which in our case is represented by salinity.

We anticipate that if evapo-concentration was responsible for the increase in solute concentrations in groundwater, there should be a positive relationship between the $\delta^{18}O$ and salinity. Our results show that in the groundwater in the Cuvelai-Etosha Basin, high salinity does not necessarily correspond with high $\delta^{18}O$ (Fig. 4); Even groundwater with moderate salinity in the Kuiseb Basin is not associated with higher $\delta^{18}O$ either. The poor relationship between salinity and the $\delta^{18}O$ is evidence that salinity is mostly due to water-rock interaction or perhaps mixing of fresh groundwater with more saline groundwater.

4.3 Elemental Composition

The order of the abundance of cations in the Kuiseb is Na > K > Ca > Mg and Na > Mg > Ca > K for Cuvelai-Etosha Basin. The marked differences in relative abundance and absolute concentrations in the major cations is due to differences in basin geology and the extent of water-rock interaction. The order of metal abundance is Fe > Al > V > As > Zn for the Kuiseb Basin and Al > Fe > V > As > Zn for the Cuvelai-Etosha Basin. We also attribute the major differences in metal composition between the two basins to geology.

We used Al and As compared to World Health Organisation (WHO) and Water Resources Management Act (Re-

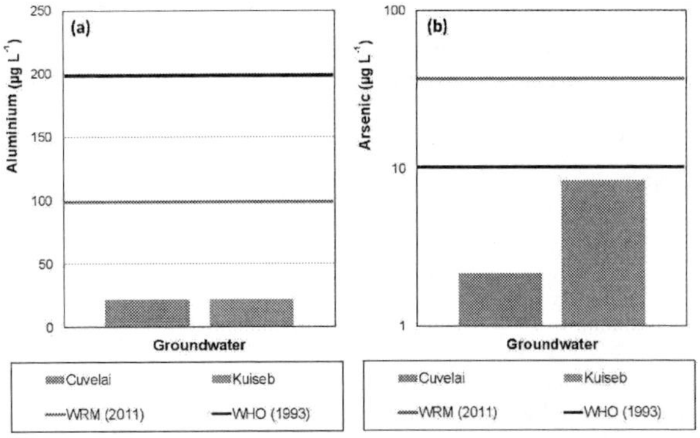

Figure 6. **(a)** Aluminium and **(b)** Arsenic levels in the Cuvelai-Etosha Basin and Kuiseb Basin. WHO → World Health Organisation and WRMA → Water Resources Management Act.

public of Namibia, 2013) standards to assess the differences between the groundwater in the Cuvelai-Etosha Basin and Kuiseb Basin (Fig. 6). Aluminium mobility and subsequent transport within the environment is influenced by chemical speciation, hydrological flow paths, water-rock interactions and the composition of the surface and aquifer materials (WHO, 1997). Arsenic in drinking water is a human carcinogen (International Agency for Research on Cancer, 1988). Humans may be exposed to arsenic in water from wells drilled into arsenic-rich ground strata or in water contaminated by industrial or agrochemical waste.

The results of our investigations show that Al is below the WHO and WRMA standards (Fig. 6a). Arsenic also has lower concentrations in groundwater compared to the WHO and WRMA Standards. Nevertheless, As has higher concentrations in the Kuiseb Basin compared to the Cuvelai-Etosha Basin. The fact that despite the marked differences in the salinity, the As concentration is much higher in the Kuiseb Basin indicates that water-rock interaction is the cause of poor groundwater quality in the Cuvelai-Etosha Basin.

5 Concluding Remarks

The groundwater in the Cuvelai-Etosha Basin has poor water quality evidenced by higher groundwater salinity classified as brackish to saline than groundwater in the Kuiseb Basin. There are higher levels of evaporation observed in more enriched stable isotopes of hydrogen and oxygen in groundwater in the Cuvelai-Etosha Basin compared to the Kuiseb Basin. The high salinity in groundwater in the Cuvelai-Etosha waters is due to water-rock interaction and less by evapo-concentration and mainly due to water rock interaction in the Kuiseb Basin. The concentration of metals (e.g., As) did not exceed the World Health Organisation and Water Resources Management Act standards which we attribute to water-rock interaction and minimal anthropogenic activities.

Data availability. The data for this article is publicly accessible from ResearchGate – https://doi.org/10.13140/RG.2.2.33777.58727.

Competing interests. The authors declare that they have no conflict of interest.

Special issue statement. This article is part of the special issue "Understanding spatio-temporal variability of water resources and the implications for IWRM in semi-arid eastern and southern Africa". It is a result of the IAHS Scientific Assembly 2017, Port Elizabeth, South Africa, 10–14 July 2017.

Acknowledgements. The authors acknowledge financial and in-kind support by the UNESCO Chair on Sustainable Water Research for Climate Adaptation in Arid Environment, Gobabeb Training and Research Centre (GTRC), National Council on Research Science and Technology (NCRST) and the Namibia Water Corporation Ltd (NAMWATER).

Edited by: Hodson Makurira
Reviewed by: Tobani Ngonidzashe and two anonymous referees

References

Christelis, G. and Struckmeier, W. (Eds): Groundwater in Namibia: an explanation to the Hydrogeological Map, 2nd Edition, 132 pp., available at: http://www.bgr.bund.de/EN/Themen/Wasser/Projekte/abgeschlossen/TZ/Namibia/groundwater_namibia.pdf?__blob=publicationFile (last access: 23 February 2018), January 2011.

Clark, I. D. and Fritz, P.: Environmental isotopes in hydrogeology, CRC press, Boca Raton, Florida, 1997.

Craig, H.: Isotopic variations in meteoric waters, Science, 133, 1702–1703, pmid:17814749, 1961.

Elliot, T.: Environmental tracers, Water Journal, 6, 3264–3269, https://doi.org/10.3390/w6113264, 2014.

Falke, M.: Kuiseb Basin water resources management project, Department of Water Affairs and Forestry, Windhoek, Ministry of Agriculture Water and Forestry, 2008.

Fritz, P., Drimmie, R. J., Frape, S. K., and O'Shea, K.: The isotopic composition of precipitation and groundwater in Canada, in: Isotope Techniques in Water Resources Development, 539–551, IAEA, Vienna, Austria, 1987.

Gardiner, M., Matros-Goreses, A., Roberts, C., and Seely, M.: Site Characterisation for Kuiseb Riparian Ecosystems, Desert Research Foundation of Namibia and Gobabeb Training and Research Centre, Windhoek, 2006.

Hamutoko, J. T., Wanke, H., Koeniger, P., Beyer, M., and Gaj, M.: Hydrogeochemical and isotope study of perched aquifers in the Cuvelai-Etosha Basin, Namibia, Isotopes in Environmental and Health Studies, January 2017.

Harbeck, G. E.: Effects of salinity on evaporation. Washington: United States Government Printing Office, available at:

https://pubs.usgs.gov/pp/0272a/report.pdf (last access: 16 January 2016), 1955.

Heyns, P. and van Vuuren, O.: Hydrology and geohydrology specialist report, in: SAIEA 2010. Strategic Environmental Assessment of the Central Namib Uranium Rush, Unpublished report for the Ministry of Mines and Energy, 2009.

IAEA: Origin of salinity and impacts on fresh groundwater resources: Optimisation of isotopic techniques, Water Resources Programme, Working Materials, 2006.

International Agency for Research on Cancer: Monograph on the carcinogenic risks to humans. Man-Made Mineral fibres and Radon, IARC Scientific Publications, 43, 173–259, 1988.

Kaseke, K. F., Wang, L., Wanke, H., Turewicz, V., and Koeniger, P.: An Analysis of Precipitation Isotope Distributions across Namibia Using Historical Data, PLoS ONE, 11, e0154598, https://doi.org/10.1371/journal.pone.0154598, 2016.

Lewis, F. M. and Walker, G. R.: Assessing the potential for significant and episodic recharge in southwestern Australia using rainfall data, Hydrogeol. J., 10, 229–237, 2002.

McGuire, K. and McDonnel, J.: Stable isotope tracers in watershed hydrology, in: Stable Isotopes in Ecology and Environmental Science, second edition, 334–375, Oxford: Blackwell publishing Ltd., available at: http://udel.edu/~inamdar/BREG667/McGuire_isotopebook.pdf (last access: 1 March 2018), 2007.

Mendelson, J., Jarvis, A., and Robertson, T.: A profile and atlas of the Cuvelai-Etosha Basin, Windhoek, Raison and Gondwana Collection, 2013.

Miller, R.: Geology of a portion of Central Damaraland, South West Africa/Namibia, South West Africa Series 6, Department of Mines – Geological Survey, Republic of South Africa, 1980.

Msangi, J. P.: Managing water scarcity in Southern Africa: Policy and strategies, in: Combating water scarcity in Southern Africa, 21–41, Springer Netherlands, 2014.

Prada, S., Cruz, J. V., Silva, M. O., and Figueira, C.: Contribution of cloud water to the groundwater recharge in Madeira Island: preliminary isotopic data, 5th International Conference on Fog, Fog Collection and Dew Münster, Germany, FOGDEW2010-97, July 2010.

Pruszkowski, E. and Bosnak, C. P.: The Analysis of Drinking Waters by U.S. EPA Method 200.8 using the NexION 300x/350X ICP-MS in Standard and Collision Modes, Shelton, PerkinElmer, Inc., 2012–2014.

Republic of Namibia: Water Resources Management Act 11 of 2013, Namibia Legal Database, https://laws.parliament.na (last access: 1 March 2018), 2013.

Schneider, G.: The roadside geology of Namibia, Berlin, Gebrüder Borntraeger, 2004.

Turewicz, V.: Stable isotopes of waters in Namibia, Germany, University of Cologne, 2013.

WHO: Aluminium. Geneva, World Health Organization, International Programme on Chemical Safety (Environmental Health Criteria 194), 1997.

Winston, W. E. and Criss, R. E.: Geochemical variations during flash flooding, Meramec River basin, May 2000, J. Hydrol., 265, 149–163, 2002.

AUTHOR INDEX